Sage on the Screen

TECH.EDU
A HOPKINS SERIES ON EDUCATION AND TECHNOLOGY

Sage on the Screen

Education, Media, and How We Learn

Bill Ferster

JOHNS HOPKINS UNIVERSITY PRESS BALTIMORE

Printed in the United States of America on acid-free paper
9 8 7 6 5 4 3 2 1

Johns Hopkins University Press
2715 North Charles Street
Baltimore, Maryland 21218-4363
www.press.jhu.edu

Library of Congress Cataloging-in-Publication Data

Names: Ferster, Bill, 1956– author.
Title: Sage on the screen : education, media, and how we learn / Bill Ferster.
Description: Baltimore, Maryland : Johns Hopkins University Press, 2016. | Series: Tech.edu : a Hopkins series on education and technology | Includes bibliographical references and index.
Identifiers: LCCN 2016007280 | ISBN 9781421421261 (hardcover : alk. paper) | ISBN 9781421421278 (electronic) | ISBN 1421421267 (hardcover : alk. paper) | ISBN 1421421275 (electronic)
Subjects: LCSH: Media programs (Education)—History. | Educational technology—History. | Web-based instruction—History.
Classification: LCC LB1028.4 .F47 2016 | DDC 372.67/2—dc23
LC record available at https://lccn.loc.gov/2016007280

A catalog record for this book is available from the British Library.

Johns Hopkins University Press uses environmentally friendly book materials, including recycled text paper that is composed of at least 30 percent post-consumer waste, whenever possible.

For Susan, Margot, and Michael

Contents

Preface

Since the days of Thomas Edison, educators have been trying to use media to offset the high cost of education. As a child of the 1960s, I certainly found the hackneyed classroom movies and filmstrips to be a welcome relief from the tedium of the typical classroom day, but it's not clear to me that the knowledge gained was worth the time and expense. The fantasy of leveraging a fixed production cost to reach an unlimited number of consumers is an enticing economic proposition, and it has been tried repeatedly as each new media format has emerged—radio, television, DVD, and, most recently, Internet-based efforts such as the Khan Academy, MOOCs (massive open online courses), the immersive learning environments of games like *Second Life*, and augmented reality tools.

Sage on the Screen explores the historical, theoretical, and practical perspectives of teaching efforts to leverage media that has been wildly successful in entertainment, but less so in education. Thomas Edison predicted in 1913 that "books will soon be obsolete in our schools," but we still see more textbooks than movies or any other form of instructional media in the classroom 100 years later.

The overall goal of this book is to explore how various media forms were created, how interrelated these forms were, and what impact they had in education. It is a follow-up to my 2014 book *Teaching Machines: Learning from the Intersection of Education and Technology*, which examined the ways teachers have used technology of all kinds to automate their instruction, with a focus chiefly on mechanical and computerized delivery vehicles.

I explore each media form by offering compelling stories about the people who promoted their use, revealing their motives and hopes in changing education using technology. These efforts are critically evaluated and common threads extracted so we can learn from over a century of work, in the hope of understanding what was effective and what was not. The focus covers the full range of learning environments, from K–20 through non–institutionally based learning.

The title of this book is a play on the old characterization of the classroom lecturer as the *sage on the stage*, as opposed to the more constructivist stance of being a *guide on the side*. The phrase was first used in 1939 in an obscure journal article about ancient pottery found in North Carolina by the archeologist Joe Caldwell: "I prefer to be the 'guide on the side,' as opposed to the 'sage on the stage.' I want my students to know that I am on their side, and I will do whatever I can to help them to learn."[1] More recently, the educator Allison King has employed the phrase as a rallying cry to introduce more constructivist pedagogy into K–12 classroom teaching.[2]

There is tremendous interest in using media in education, particularly at the college level, because the cost of education has risen beyond affordability for many people. The advent of MOOCs, where star academics make online video lectures available to thousands of students at little or no cost, has garnered attention from the mass media and popular writers such as David Brooks and Thomas Friedman.

A look back at the long history of using media in education should prove both compelling and instructional for a reader interested in modern education. This book explores a broad number of efforts, provides some overarching commentary, and offers some insight into current and future uses of this potentially valuable instructional technology, all within an interesting story about people trying to make a difference in education.

The narrative does not follow a strict chronology, as different preceding events and lines of thought influenced different forms of educational media. The discussion sometimes digresses from the primary chronology to examine these precursors, returning to the current topic to integrate its impact on the current narrative. Each chapter concludes with a summary of the impact of each genre, with some commentary as to what was and was not successful, and what factors became influences for future projects.

I recount the history with as much objectivity as possible, but I do make a conscious effort to provide my personal perspective as an educator, designer, and technologist, and to make it clear when that voice is used. I have an odd combination of pessimism about previous efforts coupled with respect for what technology can bring to society when thoughtfully introduced, and I hope to impart both perspectives.

How This Book Is Organized

"Traditional Media" (chapter 1) looks at the development of the motion picture, radio, and television. Peter Mark Roget's original insight on how we perceive motion laid the groundwork for Eadweard Muybridge's quest to capture the motion of horses galloping, which inspired Thomas Edison to successfully develop the first commercially successful motion picture. The motion picture did not revolutionize education the way Edison had predicted, but it has been popular respite for harried classroom teachers for nearly a century. The broadcast nature of radio and television divorced media from requiring a physical presence in order to be consumed. The ability to produce a lesson once and distribute it to thousands of students multiple times tugs at educators' wishes to reduce the cost of traditional classroom instruction.

"Interactive Media" (chapter 2) explores how the addition of rudimentary interactivity helped free the traditional linear media form of television by adding the ability to play the media from any point, not just start at the beginning and stop at the end. A number of short-lived innovations, such as videodiscs, opened that pathway to more effective uses of instructional media, especially when coupled with a personal computer guiding that random accessibility.

"Hypermedia" (chapter 3) examines how the ideas hinted at in Vannevar Bush's 1945 article in the *Atlantic Monthly* inspired Douglas Engelbart, Ted Nelson, and Bill Atkinson to develop systems that could navigate a learner through abstract information spaces. Apple's HyperCard made it easy to develop products that took full advantage of the new forms of interactive media becoming available—videodisc, CD-ROM, digital video—and ushered in compelling new forms of educational media.

"Cloud Media" (chapter 4) looks at the Internet's effect as a media delivery vehicle for educational media. The inexpensive and accessible nature of the Internet beckons to that same notion of scale that motion pictures instigated and later drove radio and television decades earlier. Because it is computer driven and provided on demand, the cloud has the potential to include many of the features afforded by the less flexible earlier interactive and hypermedia solutions. Cloud media can be as simple as making recorded classroom lectures available online, or the emergent MOOCs that are getting so much attention lately.

"Immersive Media" (chapter 5) investigates the more recent use of technologies that immerse the learner into realistic multimedia environments, known as virtual reality, and the emerging technologies of augmented reality. Educators used virtual reality environments, such as *Second Life*, to create "worlds" in which students explore, interact, and collaborate with one another. Augmented reality tools allow 3D media to interact with what a learner "sees" in the real world through the camera in a smartphone, tablet, or special goggles.

"Making Sense of Media for Learning" (chapter 6) sums up the earlier chapters on individual media forms and attempts to draw parallels from them. It would be difficult to point to any one and proclaim success, but there are clearly lessons to be learned from these earlier efforts—some good, and others not so much. These lessons are instructive for people who are both developing and using these new forms of instructional media.

Acknowledgments

I wish to thank the following people for their help in the research and production for this project: Malcolm Baird, Greg Britton, C. Victor Bunderson, John Bunch, Richard Clark, Dan Doernberg, Yitna Ferdyiwek, Margot Ferster, Susan Ferster, Roger Geyer, Marilyn Gilbert, Peter Gray, Ted Hasselbring, Dusty Heuston, Dov Jacobson, Kristina Hooper-Woolsey, Wendy Keeney-Kennicutt, Robert Kozma, Don McLean, Drew Minock, Bob Mohl, Ruth Perlin, John Ramo, Mark Schubin, Trevor Smith, David Staley, Aaron Walsh, Kathy Wilson, and the amazing staff of the Alderman Library at the University of Virginia.

Sage on the Screen

1

Traditional Media

A horse is a horse, of course of course,
and no one can talk to a horse of course.

Mr. Ed, 1961

For almost two centuries, media has been on an unremitting path from the physical to the abstract. Before video clips from YouTube appeared magically from out of the ether and onto our mobile phones, leaving no physical clues as to how they got there, we saw a procession of tangible devices whose inner workings were much more visible to the naked eye.

Arthur C. Clarke, author of *2001: A Space Odyssey* and a shrewd observer of technology's impact on humanity, once wrote: "Any sufficiently advanced technology is indistinguishable from magic."[1] That bar slowly rises with each innovation, and yesterday's magic is today's general knowledge. By looking at some of the more observable innovations and ideas that led to the media that is everywhere today, we can demystify it, better understand how it works, and one hopes use it more effectively for learning.

This chapter explores the evolution of more traditional analog (non-digital) media types: the motion picture and its forerunners, radio, television, and recorded media. This historical viewpoint can help us to understand the basic processes that underpin even our most sophisticated digitally encoded 3D content delivered over the Internet in ultra-high-definition resolution and multichannel directional sound. These processes are more visible, mechanical, and explainable than their digital progeny, and they provide a clearer picture for understanding the very same experience now presented to us through invisible bytes.

Persistence of Vision

Our eyes allow us to see the natural world in constant and continuous motion. To capture that moving world using technology, we need to divide continuous time into distinct periods, create a likeness of each slice of time, and store all those slices in units we call *frames*. Motion pictures and video are made up of a large number of these discrete frames, shown in rapid succession to simulate the motion we observe in the real world. Slight changes in the frame-to-frame image cause us to perceive this as motion, provided that the changes are contiguous and quick enough—around 10 to 20 frames per second. This phenomenon is why movies and video seem so real to us. Known as the *persistence of vision*, it was first discovered in the beginning of the nineteenth century by a most unlikely person.

Peter Mark Roget (of *Thesaurus* fame) was a successful London doctor (fig. 1.1) who lectured extensively on animal physiology and other biological topics. By all accounts, he was a most melancholy man with a long

Figure 1.1. Portrait of Peter Mark Roget by Thomas Pettigrew, 1843. Courtesy of the National Institutes of Health

family history of mental illness who self-medicated by completely immersing himself in his work. Like many men of his time, Roget had wide-ranging interests. Outside of his own field of physiology, he wrote more than 100 papers in a number of other disciplines and was an active participant in London's premiere showplace of scientific thought, the Royal Society. He also invented a number of mechanical and scientific devices, including the addition of the logarithmic number scale to the calculating tool of his day, the mechanical slide rule.[2]

On a late November morning in 1824, Roget happened to peer out through the basement window of his London house. He observed a horse-driven cart that was passing on the street. As he viewed the cart through the slats of the blinds, what struck him was that the spokes on the cart's wheels appeared to be curved rather than straight. Roget rushed outside and paid the carriage driver a shilling (about $1 today) to continually drive back and forth while he carefully observed from his window. To explain the perceived curvature, the doctor hypothesized, the retina sees a series of still images, when interrupted, as one continuous scene because it retains traces of the previous view. He immediately began to compose a paper accompanied by a mathematical analysis of the phenomenon (fig. 1.2).[3]

On December 9, 1824, he read this paper, "Explanation of an Optical Deception in the Appearance of the Spokes of a Wheel When Seen through Vertical Apertures,"[4] to the prestigious Royal Society in London to much fanfare. Roget himself never called this phenomenon "persistence of vision," but moving picture and media theorists have subsequently used the phrase to explain why watching a series of still images in rapid succession

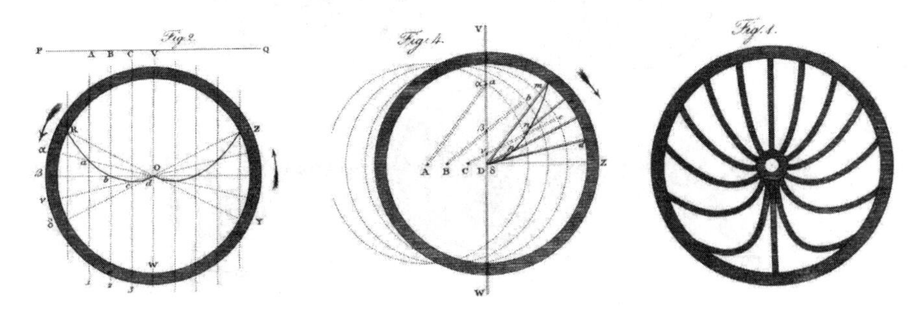

Figure 1.2. Illustrations from Roget's 1824 paper.

appears to us as motion, and each still image is interpreted as a part of a larger pattern that is continuously unfolding.

Unfortunately Roget's explanation of why we view still images as motion is not a true reckoning of the phenomenon he so deftly observed. The current thinking in neuroscience is that persistence of vision occurs from a complex interaction between maps of motion contours, rather than by comparisons of fully formed images. These motion maps are created in an area of the brain called the MT, which specializes in motion detection. The MT reacts to changes in brightness contours: motion is perceived not by comparing a previous image from the eye to the current one, but by detecting a change in the previous abstract decoding of the contours of images.

The publication of Roget's paper instigated a flurry in the invention of toys that took advantage of the persistence of image phenomena. One of the first was the *thaumatrope* (from the Greek *thauma,* "magic," and *trope,* "something that turns"), invented by another physician, Dr. John Ayrton Paris, who demonstrated it that same year at the Royal College of Physicians. The thaumatrope was a disc with two strings attached. The disc featured an image on each side, such as a monkey by itself on one side and an empty cage on the other (fig. 1.3). When spun fast enough by twisting the strings, the two images would merge together, making it appear that the monkey was now in the cage.

Thaumatropes and other similar devices that took advantage of this phenomenon were popular toys in the Victorian era for children and adults alike. Almost every pair of images was tried that made sense when blended

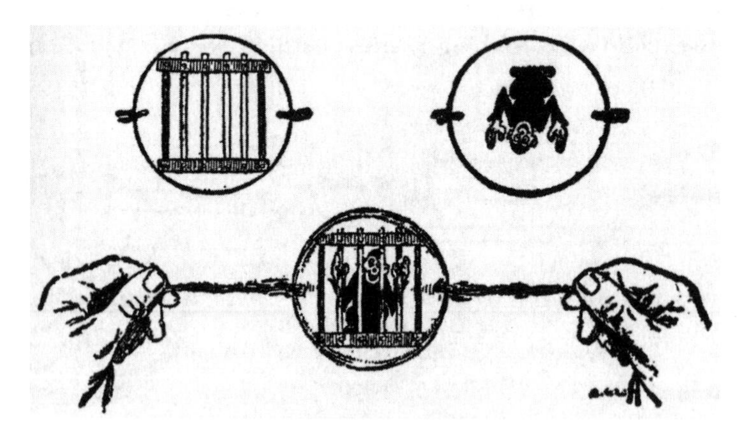

Figure 1.3. Thaumatrope, invented by John Hershel in 1825.

together: birds in cages, hunting scenes, and the ever-popular scenes depicting a horse and rider jumping a fence or galloping.[5]

But for a Horse . . .

Living in Virginia's horse country, I am reminded daily of the effect these graceful creatures have on people. Although my eyes tend to glaze over at cocktail parties when horse people talk endlessly about their equine pursuits, it's hard to miss the passion that people have for horses. In fact, the history of the motion picture owes its own debt to equine culture, and to one horseman in particular. That horseman was Leland Stanford (fig. 1.4), railroad magnate, former California governor, and future benefactor of the university that bears his name.[6]

In his day, Stanford's name would come to mind as an archetypical "robber baron." Stanford was born to a wealthy farmer and businessman near Albany, New York, in 1824, and his appetite for business arose at an early age. As a boy, he and his brothers gathered chestnuts and sold them for $5 a bushel (a bit over $100 in today's dollars), claiming toward the end of his career that it was the "most successful speculation in my life."[7]

In 1844, Stanford left his home in Albany to study at Cazenovia Seminary, a Methodist school near Syracuse, New York. He had no real religious interests and intended to study the law upon his graduation. As was the custom for legal training before formal law schools were popular, Stanford began to "read law" with a well-regarded legal firm in Albany, and he was admitted to practice law in the State of New York in 1848.[8]

The discovery of gold in California prompted Stanford's five brothers to try their luck in the gold fields. When a fire burned down his law office in 1852, Leland Stanford left Albany for Sacramento to join them in a fit of gold-rush fever. The brothers quickly gave up digging for gold, opting instead for the more lucrative tact of selling supplies to miners. They established two highly successful Stanford Brothers stores in the growing boomtown of San Francisco.

Stanford expanded his interests from business toward public service and politics in small steps. He first became a trustee of the Sacramento Public Library and later became active in the emerging Republican Party. In 1859, he ran and lost in a bid for California's treasurer and was the failed Republican nominee in the governor's race. Riding Abraham Lincoln's coattails of popularity, Stanford was elected governor of California in

1861, only four months after the attack on Fort Sumter that initiated the Civil War. Always a vigorous opponent to slavery, Stanford kept the state of California firmly within the Union during his two-year term as governor.

Stanford's greatest wealth came from his founding with a group of wealthy industrialists snidely known as the "Gang of Four" of the Union Pacific Railroad in 1861. They embarked on the exciting venture to create a vast railway network by building huge expanses of track through an unusually large number of mergers and acquisitions of other railroads, so much so that Stanford earned the nickname "The Octopus" for his voracious appetite for smaller railroad companies. In 1869, Stanford officiated at the creation of the First Transcontinental Railroad in Utah, driving in the ceremonial last spike that united the eastern and western railways into a national railroad system.[9]

Stanford had wanted to impart to Leland Jr., his 8-year-old son, the same appreciation of the farm lifestyle that he had enjoyed in Albany during his own youth, so he purchased a large amount of land in Palo Alto, just 30 miles south of San Francisco. A newspaper reporter asked during an interview late in his life which of the many professions he wished to be known by. Stanford said, "Well, I think you can put me down as farmer."[10] In the end, Stanford owned more than 8,400 acres on the beautiful California peninsula that would one day become the center of technology and the home of Stanford University.

Stanford's work as the prototypical monopolist eventually proved too stressful, and he turned to gentleman farming as a way to relieve some of that pressure. Not content to passively enjoy the country air, the industrialist brought his characteristic energy and enthusiasm to the breeding, training, and racing of horses. His farm, known as the Palo Alto Stock Farm, soon became an enormous laboratory dedicated to scientific methods to improve the speeds of racehorses, in particular, trotters. In 1870, he bought a trotter named Occident that would captivate speed records and help develop what became known as the Palo Alto System for training racehorses. Occident would later play a key role in the development of the motion picture.[11]

Unsupported Transit

There was a great debate among horsemen of that period as to whether all four of a horse's legs ever left the ground at the same time as they galloped. The action happened so fast that it was impossible to see the truth with the naked eye, and people from all walks of life had strong opinions for and against what they termed "unsupported transit." As a newcomer to the world of horses, Stanford took a strong stance that unsupported transit did indeed occur, and he put his entrepreneurial verve toward empirically proving it. The railroad magnate met and befriended an unlikely ally in his quest to prove its existence—an émigré from England and California's most famous photographer—to settle the question using the newly emerging field of photography. If he were able to record an image of all four legs of the horse off the ground, the controversy would be resolved.[12]

If Leland Stanford filled the stereotype of a tycoon—a portly, well-dressed, and manicured gentleman propped up with a pearl-covered walking cane—Eadweard Muybridge looked like a wizard from Middle Earth (fig. 1.5). A thin, spry man with penetrating blue eyes, long flowing white hair, and an equally long white beard, Muybridge looked decades older than the 49 years of age he was at the time. And he was a conjurer in the newly emerging craft of photography, which magically depicted reality through black boxes, vats, and mysterious potions.

Figure 1.4. Portrait of Leland Stanford, around 1890.

Figure 1.5. Portrait of Eadweard Muybridge. Courtesy of *Popular Science Monthly*

Born Edward Muggeridge during 1830 in the southwest suburbs of London, the son of a coal merchant, Muybridge emigrated from England to San Francisco when he was 25 years old. He had a peculiar penchant for changing the spelling of his name every few years by moving a vowel here and there, and otherwise switching consonants around. After a number of intervening variations, including Helios, he became known as Edward Muggridge, then Edward Muygridge, and finally Eadweard Muybridge. And he had as many careers as he had spellings of his name: book salesman, inventor, capitalist (he started a failed mining company), and finally an artist. In time, he became the most successful photographer in San Francisco through his beautiful images of Yosemite and the American West.

The wizard and the octopus met in Sacramento while Muybridge was taking photographic surveys for the US government.[13] Stanford and Muybridge formed an unlikely but close friendship over time. They weren't exactly peers, but there was more to their relationship than simply employer and employee. It might be better described as classical artistic patronage or an artist-in-residence. Muybridge was a frequent guest at Stanford's expansive mansion in San Francisco's exclusive Nob Hill neighborhood, and the two men were often seen rapt in deep conversation. They were clearly friends and respected one other, but Stanford's money ultimately ruled the day.

In 1872, Stanford offered Muybridge $2,000 (about $45,000 today) to photograph a horse in motion, but Muybridge was reluctant to try. The state of the art in photography at the time required long exposure times, typically accomplished by removing the camera's lens cap and replacing it after some number of seconds. Muybridge knew that to capture the desired moment, an exposure time of a small fraction of a second would be required. But Stanford was not to be deterred, and Muybridge was ultimately persuaded to try.

That May, at the Union Race Track in Sacramento, Muybridge prepared the "set" by placing white linen sheets on the ground so as to reflect the most light possible. Stanford's horse Occident was trained to trot across the sheets without spooking. After two days of attempts, with Muybridge rapidly opening and closing the lens cap by hand, it became clear that some other mechanism would be needed. He then contrived a spring-driven mechanical shutter that could freeze the action at an exposure of 1/500 of a second. Muybridge was then able to capture a blurry shadow of Occident with all four feet off the ground. Unsupported transport was most likely true, but clearer proof would be needed to settle things once and for all.[14]

That proof would need to wait a few years while Muybridge attended to some legal troubles in his personal life. In 1868 he had married Flora Downs, and the following year the couple had a son, whom they named Florado. Unfortunately Muybridge was not attentive to the marriage, and his 21-year-old wife developed eyes for one Major Harry Larkyns, a handsome literary critic. An intercepted letter from Flora to Larkyns in 1874 led Muybridge to suspect that Florado was the product of that union rather than his own, and he confronted Larkyns at a miner's saloon called the Yellow Jacket. The photographer calmly walked up to Larkyns and said, "My name is Muybridge, and I have received a message from my wife." Muybridge then raised a gun and shot him point-blank in the chest. He politely apologized to the women in the saloon, saying "Sorry for the disturbance."[15] Muybridge was promptly arrested and tried for Larkyns's murder.

The event was the O. J. Simpson trial of its day, having all the ingredients for good drama: sex, betrayal, passion, a handsome victim, and a well-known and unusual defendant. The recent introduction of the telegraph meant that the entire nation followed the trial in real time. Nearly a century later, the composer Philip Glass's *The Photographer* contained a libretto that used parts of the trial's transcript. Even though he defended himself by using an insanity plea, Muybridge was ultimately acquitted of the murder on the grounds of "justifiable homicide." He then traveled to South America on a nine-month photographic excursion to let things settle down.[16]

Toward the end of the nineteenth century, Muybridge was not the only person fascinated by capturing motion. Étienne-Jules Marey was a respected physician and professor of natural history at Paris's Collège de France. As a boy, he was something of a mechanical genius, to the point that his mother once described him as having "brains in his fingers."[17] He was a prodigious author of articles in physiology and an inventor of a number of revolutionary medical instruments, including the cardiogram and an apparatus to measure pulse rates.

Marey also had a keen interest in recording the motion of animals, and he invented delicate instruments to capture the flight of insects and birds (fig. 1.6). He used photography to capture these images; in 1873 he published a stunningly beautiful book of these photos titled *La Machine animale, locomotion terrestre et aérienne*[18] (The Animal Machine, Terrestrial

Figure 1.6. Photograph of flying pelican taken by Étienne-Jules Marey, 1882.

and Aerial Locomotion). Around this time, Muybridge was aware of Marey's work and owned a copy of his book. Stanford most likely learned of Marey from an 1874 article in *Popular Science*[19] about studying the gait of horses, a topic that was clearly of interest to Stanford, particularly because Marey, too, believed in the unsupported transit theory through his own work on horses.

Capturing Time

Muybridge's crime, trial, and acquittal apparently had little effect on his relationship with Stanford, and in 1876 Stanford used the *Popular Science* article to cajole Muybridge into providing clear photographic evidence of unsupported transport. By the time they were finished, the project had cost over $50,000 (close to a $1 million today).[20] Muybridge needed a much faster shutter to capture a horse trotting at 38 feet per second, so he worked with electrical engineers that Stanford provided from the Central Pacific Railroad to develop a camera shutter that was electrically operated and could make the extremely short exposures required.[21]

Muybridge devised a new plan to capture the exact time when the horse might be fully aloft. Having previously visited Marey's laboratory in Paris, Muybridge was aware of Marey's use of a single camera to record motion but chose to take a more elaborate approach. Instead of using just one camera, and hoping to capture the precise time of the fleeting event, Muy-

Figure 1.7. Apparatus for filming a galloping horse, 1881.

Figure 1.8. Sequence of a horse galloping, by Eadweard Muybridge. For an animated GIF of this image in motion, see www.viseyes.org/horse.gif.

bridge bought a dozen expensive cameras and lenses and laid them out in even intervals across the set at Palo Alto Stock Farm. Each camera's shutter was connected by a silk thread that would break as the horse strode through the set, triggering the capture of an image.

On June 15, 1878, Muybridge invited members of the press to watch Stanford's horse gallop across the set he had prepared. Occident triggered each of the 12 cameras, one after another, making a sound like a drum roll (fig. 1.7). Over the next 20 minutes, Muybridge developed the glass plates in a lab he had constructed next door. He then delivered a dozen perfectly crisp images of the horse in motion—including one that showed the equine with all four feet clearly in the air (fig. 1.8). The validity of unsupported transport was now undeniably proven.[22]

Figure 1.9. Studies in zoopraxography arranged for the zoopraxiscope, 1893.

Figure 1.10. Zoopraxiscope, circa 1894.

The idea that motion in the world could be captured at any point was as transcendental as photography had been a mere 50 years earlier. Before photography emerged in the early nineteenth century, artists mediated visual representations of the world, and they consciously chose what and how to depict them. Photography shortened that interpretive chain by using technology, and Muybridge's application of photography to capture motion paved the way for these slices of time to be combined so as to reconstruct the motion they discretely captured.

At a lavish party at Stanford's Nob Hill mansion on January 16, 1880, Muybridge demonstrated the first projected moving picture to a rapt audience of San Francisco's elite. To show two seconds of a horse in motion, he modified a lantern projector that was typically used to project still images painted on glass slides. Called the *zoopraxiscope*, it employed a mechanism that rapidly spun a glass disc with 12 of his frozen images of a horse galloping around its circumference (fig. 1.9) between the gas-fed light source, a shutter, and a lens (fig. 1.10). The audience, which included newspaper reporters, was mesmerized. The world was one step closer to motion pictures.[23]

Motion Pictures

Muybridge's two seconds of animation were groundbreaking and exhilarating to watch, but having a small number of images rotating around a disc necessarily limited the amount of time that could be shown. Many

Figure 1.11. Edison and his phonograph, 1890. Drawing by Anton Nordgren

innovators looked for alternatives to increase the length of the projection, including an American with a voracious appetite for invention known as the Wizard of Menlo Park.

Thomas Alva Edison needs no real introduction for most readers (fig. 1.11). He was America's preeminent inventor during the late nineteenth and early twentieth centuries, with over a thousand patents to his name for such seminal inventions as the stock ticker, electric light, phonograph, microphone, and—of special interest to us here—the moving picture industry.

Unlike other inventors and academics such as Muybridge and Marey, who had dabbled in moving pictures, Edison had not only the drive to invent the basic technology, but also the vision to promote those inventions within whole ecosystems. Just as Apple's Steve Jobs would later do with the iPod and iTunes, Edison created the technology for both recording movies (the Kinetograph) and playing them back (the Kinetoscope), and he also actively produced movies in his studio in New Jersey and created distribution networks for their popular dissemination. (Edison had previously done the same with his wax-cylinder phonographs.)

In 1887, he charged an assistant, William Dickson, to begin assembling a mechanism that could record microimages onto a drum using a Zeiss microphotography system, much like his early phonographs. Unfortunately Dickson did not initially make much progress on the project, because at the time attention was needed for issues more pressing to the Edison Manufacturing Company. In time, however, he would become one of the more prolific directors of early motion pictures for Edison.

William Kennedy Laurie Dickson was a tall, dapper, and mustachioed Scotsman (fig. 1.12) who joined Edison's company in 1879 after the death of his wife in Virginia. Many film scholars, in particular the film historian Gordon Hendricks, believe that Dickson was the true inventor of Edison's motion picture projects, and they accuse Edison of taking undue credit. But the truth lies somewhere in between. Dickson directed the majority of Edison's early films and acted in a number of them. He was an avid photographer and he contributed considerable skill in that part of the invention, but the motion pictures were equally mechanical and photographic in nature, and Edison's talents there were legendary.

Meanwhile, during the winter of 1888, Eadweard Muybridge was on a speaking tour promoting his zoopraxiscope to thrilled audiences across

Figure 1.12. William Dickson playing a violin into a recording horn, 1894.

the country. On February 23, he performed in Edison's hometown of Orange, New Jersey, and Edison and Dickson reportedly attended the lecture. Two days later, they invited Muybridge to visit Edison's West Orange laboratory. The two wizards talked about combining Edison's phonograph with Muybridge's zoopraxiscope to "reproduce simultaneously in the presence of an audience, visible actions and audible words."[24]

Edison had been thinking about combining moving pictures with sound for almost a decade, but he had been preoccupied with his many other endeavors until 1887. (Ironically, during an interview in 1930, a year before his death, he mentioned that "talkies spoiled everything," adding, "There isn't any good acting on the screen. They concentrate on the voice now and have forgotten how to act. I can sense it more than you because I am deaf."[25]) This meeting encouraged Edison to continue to explore motion pictures, but apparently not with Muybridge. The following year, Edison visited Paris to see the 1889 World's Fair, where the Eiffel Tower made its debut. He was squired about town as a celebrity among the Parisian scientific elite. During his visit, he met with Muybridge's friend Étienne-Jules Marey, and he saw the work Marey was doing in capturing motion on film. This was the final stimulus for Edison to begin the process in earnest.

In particular, he was impressed with Marey's *chronophotographic gun*, which resembled an elephant rifle. The chronophotographic gun was able to capture 12 frames per second of motion onto flexible strips of photographic paper that were coiled around the inside circumference of the round magazine sitting on top (fig. 1.13). Unlike Muybridge, who had dismissed the

Figure 1.13. Marey's chronophotographic gun. Courtesy of David Monniaux

value of Marey's cameras, Edison had an epiphany that his and Dickson's tack of using a phonograph-style drum might be the wrong approach to recording images. Almost immediately upon his return to America, he redesigned his camera to record images on strips of cellulose film.[26]

Edison returned to Orange on October 6, 1889, and two days later filed his first motion picture caveat with the US Patent office. A *caveat* is a statement by an inventor of his intent to file a patent and to establish legal precedence. Edison began with this provocative statement: "I am experimenting with an instrument which does for the Eye what the phonograph does for the Ear."[27] In all, he filed four caveats before their invention was revealed some 18 months later.

Edison was able to improve upon Marey's recording mechanism by adding a series of punched holes to better move the paper in front of the camera's lens, a trick derived from his earlier invention of the ticker tape. The new device, which he called the *Kinetograph*, employed a clever mechanism that made it possible to advance accurately measured jumps of 35-mm-wide paper. Using a hand crank, Edison's camera could expose 50-foot strips of George Eastman's new cellulose film that could record around 1,000 frames. This translated to a recording length of somewhere between 20 and 60 seconds, depending on how fast the operator wound the crank.[28]

To play back these recordings, Edison and Dickson devised a large cabinet containing the filmstrips wrapped on a series of rollers (fig. 1.14). Atop the cabinet was a peephole containing a lens-rotating shutter mechanism

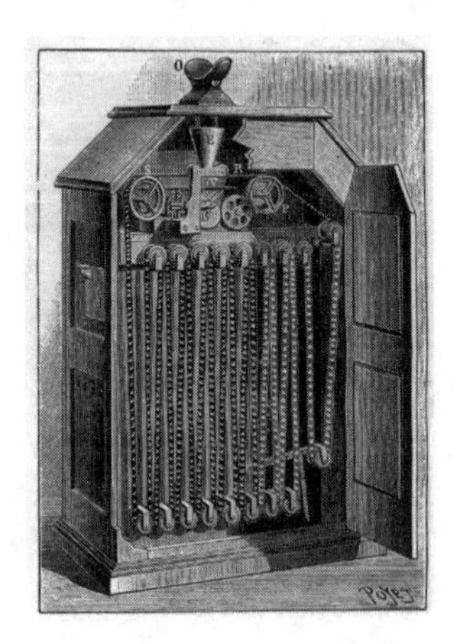

Figure 1.14. William Dickson's Kinetoscope, circa 1895.

that presented the individual frames on the filmstrips as they whizzed by on the rollers below. The Kinetoscope debuted to an impressed audience on May 20, 1891, at a press event during the annual meeting of the Women's Clubs of America Convention in Orange. Aside from the reporters, 147 clubwomen were present, including Edison's wife, Mina.[29]

Looking to provide a space to create movies that the Kinetoscope would play, Edison and Dickson built the first motion-picture studio in 1893. They called it Black Maria, and it was covered in black tar paper to keep out light. In keeping with our horse theme, it was named after the slang term used to describe the black horse-driven paddy wagons used by the police of the day to transport prisoners. The back roof could be opened to let in sunlight for filming, and the whole building could be rotated like a railroad turntable to capture the sun in its varying directions (fig. 1.15). Dickson brought in performers to be filmed in various scenes, including some that were off-color for the era: scantily clad women dancing, boxing matches, cockfights (which were outlawed at the time), and even performers from Buffalo Bill's Wild West Show.[30] And, proving the historical precedence of Internet cat videos, there were Professor Welton's Boxing Cats (fig. 1.16; Google "boxing cats" to see the actual movie).

Figure 1.15. Exterior view of Edison's Black Maria film studio. Illustration by von E. J. Meeker, 1894

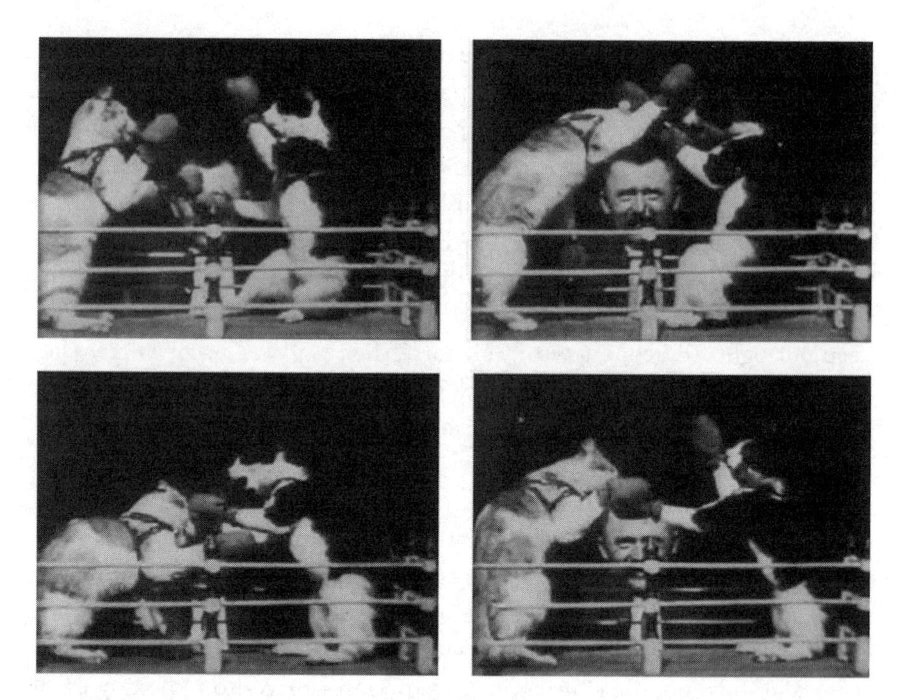

Figure 1.16. Before cat videos on YouTube, there were Professor Welton's Boxing Cats. Photo by W. K. L. Dickson, 1894

Edison's company quickly began commercializing its new invention. On April 14, 1894, the Holland brothers opened the first of many Kinescope parlors in Manhattan, at 1155 Broadway (at 27th Street). There were ten Kinescopes: five on each side with operators standing in between them. Lines of patrons streamed past, depositing quarters (worth about $5 today) and viewing the short films shot at the Black Maria.[31]

Edison began selling Kinetoscopes bundled with four films to potential exhibitors for $250 (around $6,000 today), and the company offered individual films from a catalog of 60 titles for $10 to $15 each. There were 15 films of people dancing, 9 of people in combat of one sort or another, and 31 of a general nature.[32] Exhibitors opened parlors with multiple Kinetoscopes, known as *Nickelodeons*, because a single film cost only a nickel to see. Kinetoscopes rapidly grew in popularity; only three years after their introduction, there were up to 10,000 parlors showing films.[33]

Projecting Movies

Compared with Muybridge's zoopraxiscope, the images in the Kinetoscope were smoother, higher resolution, and lasted for minutes rather than seconds. But the apparatus offered only a private viewing experience in which the watcher peered down an eyepiece at the top of the massive box. The zoopraxiscope, in contrast, could be enjoyed by many people at the same time. Dickson began to work on projecting motion-captured images from Muybridge's book by reconstructing a device he had read about in the November 1889 issue of *Scientific American*[34] called the *tachyscope.* Edison objected to the small size of the images, which was only 4 inches by 5 inches, and also its terrible flicker. After a few months, he ordered Dickson to stop experimenting with projection. Dickson apparently took great umbrage to this, and after leaving Edison's employment in 1897 he started a company (that proved to be unsuccessful) to produce projectors.[35]

Edison ultimately succumbed to the need for projection for his films, but instead of inventing his own, he acquired a projector developed by Thomas Armat and Charles Francis Jenkins that was first shown publicly in 1895. The duo sold the patent to Edison's company, and Armat went to work for them the following year. (Jenkins would later play an important role in the development of early television.) The *Vitascope* premiered to great fanfare at Koster and Bial's Music Hall in Manhattan on April 23, 1896, and ushered in a more social way to view motion pictures.[36] But the Vistascope was not the first machine to project motion pictures to an audience. A year earlier, French brothers Auguste and Louis Lumière showed their Cinématographe in Paris. Although their device did not enjoy the commercial success of Edison's Vitascope, the duo's filmmaking prowess pushed the boundaries of early cinema.[37]

Educational Films

Educational films pioneered the idea of "edutainment" productions that offered educational content presented in an engaging style. Even though the inventors of the motion picture originally envisioned film as an educational medium—recall that its origins were in examining animal locomotion—it was primarily used for entertainment until a large enough collection of titles specifically produced for educational purposes became available in the 1910s, and schools began to purchase projectors for classroom use.

Thomas Edison was so enthralled with the potential of educational film that in 1913 he pronounced: "Books will soon be obsolete in our schools. . . . Our school system will be completely changed in ten years."[38] His former assistant William Dickson, too, had high hopes for educational films, saying, "The advantage to students and historians will be immeasurable. Instead of dry and misleading accounts, tinged with the exaggerations of the chroniclers' minds, our archives will be enriched by the vitalized picture of great national scenes, instinct with all the glowing personalities which characterized them."[39] But like most proclamations of technology changing schools, these would not be prophetic (fig. 1.17).

The first educational films were inspired by the newsreels that ran before the start of feature films shown in movie theaters. Composed of short film clips from Pathé, News of the World, the March of Time, and most of the major Hollywood studios, newsreels documented current world events with a narrated overview. To this day, these newsreels and their outtakes provide filmmakers with a wealth of historic footage, much of which is used in modern documentaries. Early educational films used outtakes from these newsreels and castoffs from other commercial film projects to create short educational subjects. Between 1911 and 1914, Edison's company produced a number of historical films for the classroom as well a number of titles on natural and physical science.[40] In 1910, George Kleine created the first catalog dedicated to educational films, with more than 1,000 titles in 30 subjects. He was realistic about the educational nature of many of his films, saying, "In a sense, all subjects are educational but in classifying a mass of motion pictures for educational purposes, the line must be drawn about a reasonable area. . . . The word 'educational' is used here in a wide sense and does not indicate they are intended for school or college use exclusively."[41]

But the adoption of film into the classroom was hampered by the technology of film itself. In 1910, Kleine tried to get New York City public schools to adopt his educational films, but despite strong support from the board of education, there was a lack of infrastructure to project them.[42] Commercial film was large, 35 mm in width, requiring heavy and bulky projectors. The film was also expensive to use; hefty 14-inch reels holding 1,000 feet of film that played for only 11 minutes. In 1912, a smaller, 16-mm-gauge film was developed, and a number of manufacturers began making less expensive and more portable projectors to fit the new compact-sized film. The 16-mm film was also more economical, requiring only 400 feet on a smaller 7-inch reel to play the same 11 minutes as the larger-sized film.

It wasn't just the size of the film that made it hard to use in the schoolroom environment. It was also an issue of safety. During that time, motion pictures were made of an extremely flammable cellulose-nitrate (commonly known as nitrate) plastic that burned like phosphorus when lit, say by a projection lamp shining too close. Strict safety laws were quickly enacted to protect viewers, but this danger kept schools from showing

Figure 1.17. Cartoon in the *Chicago Tribune* by John McCutcheon, 1923.

nitrate-based films in the classroom. In the mid-1920s, Eastman Kodak developed an acetate-based "safety film" that was significantly safer than the older nitrate-based film, which was prone to spontaneous combustion. In 1978, a massive explosion at one of the National Archives' storage buildings in a Maryland suburb of Washington, DC, leveled the structure and destroyed over 13 million feet of historic footage.[43] Both the Library of Congress and the National Archives have strong initiatives in place to convert the millions of feet of the remaining volatile films to the more archivable safety format.

By the mid-1920s, movies were firmly entrenched in American life, and they began to be widely used in schools. The movies produced at that time tended to be from three genres: *entertainment*, *educational*, and *pedagogical* (or instructional). Entertainment films had commercial purposes and were shown in theaters, educational titles were meant to be informing, and pedagogical films provided specific instruction on subjects in an educational context, with significant overlap between those subjects. The 1922 classic film *Nanook of the North,* which "documented" the life an Inuit man in the Canadian Arctic, was as entertaining as it was educational in a general sense. It was also pedagogical in a class about ethnography.[44]

These labels would not hold up well if we were to apply the same criteria to the classical texts commonly used in classrooms today. William Shakespeare's plays were originally written for commercial gain in the Globe Theatre during the seventeenth century, but today they are educational staples in high school English classes. The distinction seems to hinge more on how they are used rather than the purpose they were originally created to serve.[45]

The production of educational films began to come into its own genre in the 1950s when encyclopedia companies and the National Film Board of Canada produced films that were fundamentally different from newsreels and Hollywood features. These films had high production values and took on distinct narrative styles to convey information other than the "stentorian" style of the typical male narrator, delivered in a monotone. Films used personas to drive the narrative arc—such as a *detective* solving a case, a *tourist* in a distant land, a *social critic* investigating a societal problem, a *prosecutor* exposing neglect or impropriety, and a *referee* presenting multiple perspectives—and these personas and situations made the films more interesting to students.[46]

Figure 1.18. Ford advertisement in *Moving Picture Age*, January 7, 1920.

The makers of projectors produced some of the first films: Kodak, Victor, Bell and Howell, AT&T (through their manufacturing arm, Western Electric), and even the Ford Motor Company (fig. 1.18). As the "talkies" began to overtake silent pictures, AT&T's production efforts, branded under the moniker Electrical Research Products Inc. (ERPI), began to produce sound films for education, and ERPI quickly rose as a leading producer of classroom films. In a bid to add some intellectual credibility to their films, in 1932 ERPI began an alliance with the University of Chicago for academic support. Antitrust pressure on AT&T in 1937 from the US Department of Justice led to the sale of ERPI to the popular Encyclopaedia Britannica, which soon became the largest producer and distributor of educational films in the world. Britannica also had a close relationship with the University of Chicago at that time; the university ultimately acquired the company in 1941.[47]

Early research on using educational films in the classroom showed positive results. A two-year study sponsored by Kodak in 1923, involving over 11,000 kids in a dozen cities across the country, found that "Students taught by film were found to excel in questions of fact, and also demonstrated superiority in explanatory and conceptual test items." These types of comparative studies were abandoned by the 1940s because they all gave the same results: educational films could convey the same content as effectively as other methods, but 10 to 20 percent faster.[48]

Radio

The march away from the physicality of film toward electrons was an important shift if images were ever to be transmitted remotely through the air like radio waves. By the mid-1920s, radio had taken the world by storm, with approximately 60 percent of American households owning a set that could receive radio programming from nearly 600 radio stations. As with photography, amateur participation in early radio was strong, from the construction of simple crystal radio receivers, with hand-wound coils, to more elaborate homebuilt vacuum-tube transmitters. It all started with the experimentation of a 19-year-old Italian man around the turn of the twentieth century.[49]

Guglielmo Marconi (fig. 1.19) began experimenting with transmitting signals using radio waves across the third-floor bedroom in his family's villa, near the northern Italian city of Bologna. His wealthy father was skeptical of his activities—until he was able to ring a bell on the ground floor from his bedroom. Signor Marconi was impressed and gave his son his full moral and financial support to refine the technology, and, over time, the younger Marconi was able to transmit these signals farther and farther.[50]

The Italian government showed little interest in Marconi's work, so his Irish mother offered to take him to the United Kingdom in search of backing. In 1896, the pair sailed to England at the request of the British Post Office, and Marconi was able to transmit a telegraph signal over a distance of 9 miles. He said later that "the calm of my life ended then."[51] The age of radio had begun, and Marconi received the 1909 Nobel Prize in physics for his pioneering work in radio.

Like many new technologies, the first uses of radio were military and for commercial ship-to-shore communication. The technology steadily developed, and Marconi's company, American Marconi, grew. A turning point came during the evening of April 14, 1912, when a young telegraph operator at Wanamaker's department store in New York City received a distress call from an ocean liner that had struck an iceberg. The ship was the *Titanic*, and the radio operator was a 21-year-old Russian immigrant

Figure 1.19. Guglielmo Marconi posing with his wireless apparatus, 1909.

named David Sarnoff, who said some years later, "The *Titanic* disaster brought radio to the front, and incidentally me."[52]

There has been some debate among historians whether this story is true, but Sarnoff would later play a pivotal role in the commercialization of radio, and also in the development of early television and videodiscs. Ironically, Marconi had been offered free passage by the White Star Line on that fateful maiden voyage of its *Titanic*, but he chose instead to travel on another ill-fated ship, the *Lusitania*, three days earlier. His daughter later said that he had paperwork to do and preferred the *Lusitania*'s public stenographer over the *Titanic*'s. In a second stroke of luck, Marconi crossed the Atlantic Ocean aboard the *Lusitania* on its last voyage before being torpedoed and sunk by a German U-boat in 1915.[53]

Sarnoff and Marconi became great friends, and the young Sarnoff held increasingly more important jobs with the company as it grew into one of the leading radio manufacturers by the mid-1920s. American Marconi was later purchased by General Electric (GE) and rebranded the Radio Corporation of America (RCA). With that purchase, GE acquired David Sarnoff, a future power player in broadcasting. His star rapidly rose within RCA as he took on more responsibility, eventually becoming its president in 1930.

Instructional Radio

It would be easy to dismiss educational radio as a failed experiment, one that tried to harness the technology of the day only to be superseded by the next best thing. But the story is a bit more nuanced. Radio turned out not to be as effective in urban schools, but it provided outreach to countless rural classrooms that, not unlike some schools in today's rural America, lacked the range of faculty able to teach a broad enough curriculum. It later provided an opportunity for students themselves to take an active role in producing instructional programs.

In the mid-1920s, schools began experimenting with using radio in a smattering of small efforts in public school systems throughout the midwestern United States, as well as in a few colleges and universities. Radio was much more accessible to teachers than film: it did not require expensive cameras, editing equipment, projectors, or the highly specialized skills that movies required. Radio transmitters and receivers were inexpensive and easy to operate, and did not require a constant supply of costly film stock to feed instruction. Unfortunately, like the educational film and video makers, university efforts proved ineffective because professors simply lectured into a microphone. It became evident that the skills of a good lecturer did not necessarily qualify one as an effective broadcaster.[54]

Benjamin H. ("Uncle Ben") Darrow (fig. 1.20), after years of missionary work, attended the agricultural extension of the Maryland State College, where in 1924 he was in charge of children's programs at WLS radio, the Sears, Roebuck–owned commercial station. At WLS, Darrow started an innovative radio show called *The Little Red School House of the Air*, which broadcast lessons in art, music appreciation, and geography. Each episode

Figure 1.20. Educational radio pioneer Benjamin H. Darrow. Courtesy of the Ohio State University Photo Archive

combined dialogue and music around a particular topic, all written and performed by students and their teachers. They rehearsed using dummy microphones at school, and then came into the studios to broadcast the programs at set times each week. *The Little Red School House of the Air* was very well received by both rural and city schools throughout its Chicago-based listening area.[55]

Darrow's time at *The Little Red School House of the Air* empowered his sense of radio's potential impact on education and led him on a Sisyphean crusade to gain support for other instructional radio programs. He saw radio as a kind of global village, where "the voice of the world becomes one neighborhood."[56] He succeeded in establishing a number of successful spinoff programs throughout the country, including the *Wisconsin School of the Air*, which broadcast for more than 40 years.

A number of researchers who studied the effectiveness of radio instruction in the 1950s generally concluded that radio afforded little or no gain over traditional teaching techniques.[57] Instruction by radio may also have suffered from its focus on a single sensory input: hearing. Even in a lecture, the visual presence of the speaker can communicate far more information than the sound of the speaker's voice alone. The facial and hand gestures, not to mention the writing on the blackboard, contribute much to the communicative experience. I am a big consumer of recorded books and listen to many podcasts each week, but there are not many times I listen without performing some other activity, such as walking or driving. Sitting in the classroom and passively listening to radio broadcasts may have been exciting at first, but ultimately it was not immersive enough of an experience for students to embrace.

Television

By the 1920s, the desire to add a visual component to radio was attracting a wide range of inventors eager to develop the "next big thing." The process proved more difficult, and a wide range of approaches were tried, mostly mechanical in nature, many using the same Nipkow disc with the multiple lenses that television pioneer John Logie Baird used later in his innovative Phonovision. The idea that the image needed to be scanned into lines and those lines converted into electrical signals was not in dispute. The question was how best to do it. The radio companies had grown into large corporations, with well-funded research laboratories that developed innovation after innovation and accumulated masses of patents to protect those inventions. But it would be a lone young inventor whose epiphany while still in high school, followed up by decades of hard work, finally developed television.

Philo T. Farnsworth's (fig. 1.21) story is almost too archetypically mythical to be true. He was born in 1906 in southern Utah in a log cabin with no electricity. His father was the son of a Mormon bishop who had been among the first settlers sent by Brigham Young in 1856 to settle Utah for the newly formed religion. Finding the land difficult to farm, Farnsworth's father moved the family to Rigby, Idaho, in 1918. Previous owners of the ranch had installed electricity furnished by a generator, which fascinated the young Philo. More importantly, in the attic the previous owners had also left behind stacks of catalogs, technical journals, and amateur radio and popular science magazines.

Figure 1.21. Philo T. Farnsworth, 1939. Courtesy of the Library of Congress

It's hard to overstate the importance the magazines of this time had on spurring innovation among independent inventors, and Philo Farnsworth was no exception. He devoured articles on crystal radio receivers as well as advertisements for radio kits, vacuum-tube transmitters, and, in particular, Paul Nipkow's rotating disc for capturing imagery. Hugo Gernsback's popular *Radio News* magazine coverage of inventions for sending images over the radio piqued Farnsworth's interest in the subject as he worked on the family's farm.[58]

Farnsworth believed that those mechanical schemes of scanning an image would never yield the resolution needed for a realistic depiction of reality. His big epiphany came in 1921 as he was plowing the hayfield atop a single-disc harrow drawn by three horses. He visualized the orderly rows he was creating as lines dividing an image. From his reading, he knew he could create those lines using electrons in a vacuum tube. In this way, he

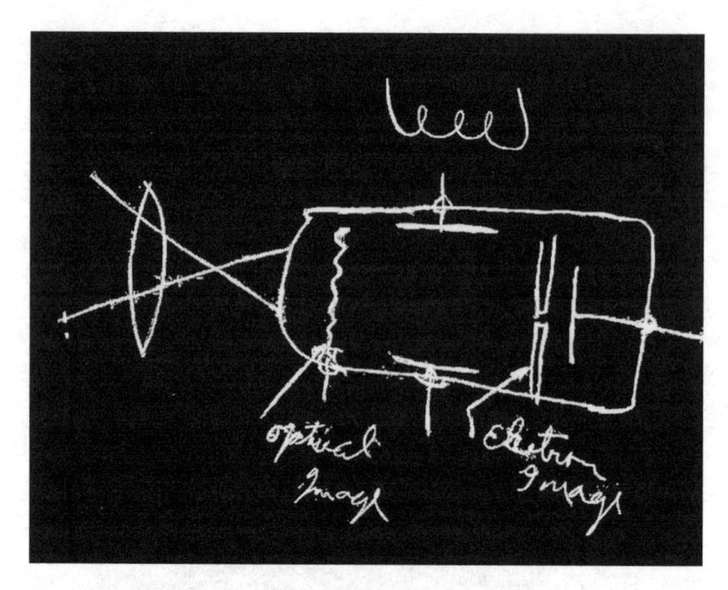

Figure 1.22. A sketch of Farnsworth's 1922 blackboard drawing. Adapted from images courtesy of the Farnsworth Archives

could eliminate the awkward mechanical means that previous scanning approaches needed with invisible but fast-moving electrons. At Rigby High School in the spring of 1922, the 15-year-old prodigy methodically drew a diagram of his idea on his favorite chemistry teacher's blackboard (fig. 1.22). A sketch of this drawing would prove most valuable nearly two decades later in patent litigation.[59]

Farnsworth would later call his invention the *Image Dissector.* He worked tirelessly for years to develop his idea, and to raise the needed capital to accomplish its creation. At the urging of one his financial backers, he moved to California and set up a makeshift laboratory in a Hollywood studio. Behind closed blinds, he worked in secret on his invention, ultimately raising suspicion from the local police that he might be distilling liquor in the era of Prohibition. Authorities raided his laboratory in 1926 and found blown glass tubes, but no alcohol. By the following year, he was making progress and was able to transmit crude images and scenes.[60]

In perfecting the design, Farnsworth used movie footage as test subjects, first of a hockey game and later a scene of a popular film actress in the 1929 film version of Shakespeare's *The Taming of the Shrew.* He played the clip so often while making adjustments to the device that one of his financial backers remarked, "Mary Pickford combed her hair at least a million times for the benefit of science and the development of television."[61]

This Time the Goliath Was Named David

Meanwhile, the radio companies used their massive research and development resources to add vision to their popular audio offerings. The major radio manufacturers—Philco, GE, Westinghouse, and RCA—all had invested heavily in television research, viewing it as a critical next phase for growth, but only RCA would develop a successful product. This was primarily because of David Sarnoff's untiring, and sometimes unethical, efforts and the work of a brilliant young Russian physicist who had just immigrated to the United States.

Vladimir Zworykin was the scion of a wealthy Tsarist family studying electrical engineering in St. Petersburg, but following the Bolshevik Revolution of 1917, he found himself a man without a country. He moved to Pittsburgh to work for Westinghouse, where researchers were investigating mechanical scanning methods for television, while simultaneously earning a PhD in physics. Over time, much as Farnsworth had intuitively known from the start, Zworykin lost faith in the mechanical approach to do the scanning in favor of directing electrons in a vacuum tube.[62]

Just as William Dickson had attempted to convince Thomas Edison that projection was critical to the success of the motion picture business, Zworykin did not believe the mechanical approach to scanning images would ultimately prove an effective solution, and he was likewise unable to convince his employers of its value. He created a prototype that used electrons, not mechanics, to scan and display images, but Westinghouse ignored his ideas. Frustrated, he made an appointment in January 1929 with Westinghouse's competitor, RCA, to pitch the idea to David Sarnoff. Intrigued, Sarnoff asked him how much he thought the project might cost. Zworykin responded that he guessed around $100,000 ($1 million today), to which Sarnoff replied, "All right, it's worth it."[63] This figure proved to be a gross underestimate; the actual cost was well over $50 million ($5 billion today).

Zworykin was making steady progress on his own electronic scanning method, but he was also well aware of Farnsworth's work. Farnsworth had been actively publicizing his efforts in the trade and popular presses in order to raise money, and both Sarnoff and Zworykin followed his developments closely. On April 16, 1930, only three months after his initial meeting with

Sarnoff, Zworykin flew to California to meet Farnsworth. Holding the young inventor's camera tube, he praised it, saying, "This is a beautiful instrument. I wish I had invented it." Just a few weeks later, on May Day, Zworykin filed a patent of his own.[64]

Both Farnsworth and Sarnoff took the filing of patents seriously. In the end Farnsworth wrote an impressive 165 patents, but RCA was a veritable patent factory, filing thousands of them. Sarnoff fed this invention factory by hiring bright university students and offering them an opportunity to change the world by paying them well and by providing well-appointed laboratories and an intellectually stimulating environment. In return, Sarnoff demanded they sign over the rights on any patents to RCA for a check in the amount of $1. An unusually prolific engineer by the name of Bill Eddy found this policy absurd, so instead of dutifully cashing his checks, he pasted them to his office wall. When the accounting department noticed the uncashed checks, a visit to Eddy's office revealed an entire wall of them. Unable to remove the checks from the wall, the accountants resorted to cutting out a section of the plasterboard and carrying it to the bank. They returned with a small stack of dollar bills, thereby consummating the contract.[65]

Zworykin made it clear to Sarnoff that the approach he was taking was similar to Farnsworth's and that RCA would need Farnsworth's patents in order to succeed. But as a matter of policy, RCA did not license other inventors' patents. After multiple failed attempts to buy the young inventor's company, Sarnoff engaged Farnsworth in a protracted and expensive legal battle to determine the first inventor of the technology, sometimes using unethical tactics. The case raged on in the courts for years. Ironically, the issue of who actually invented television hinged on one piece of evidence, a sketch of the drawing Farnsworth drew on his chemistry teacher's blackboard, to determine who had invented the television.

Finally, on October 2, 1939, the courts ruled in favor of Farnsworth, and RCA signed its first patent licensing deal in company history with him. Philo T. Farnsworth was officially crowned the inventor of television. Neither Sarnoff, Farnsworth, nor Zworykin were present at the signing, but legend has it that RCA's lead attorney, Otto Schairer, was so distraught that he had tears in his eyes when he leaned over to sign the final agreement.[66]

RCA unveiled commercial television in April 1939 at the World's Fair in New York City. That year's theme was "Building the World of Tomorrow," and Sarnoff spared no expense in making sure television was a part of that

Figure 1.23. David Sarnoff introducing television at the 1939 World's Fair in New York. Courtesy of the David Sarnoff Library.

world. People streamed past television cameras, and their images appeared on the many television receivers in the massive 9,000-square-foot RCA pavilion. Speaking behind a podium draped with his company's logo (fig. 1.23), Sarnoff said, "Television is an art which shines like a torch of hope to a troubled world. It is a creative force which we must learn to utilize for the benefit of mankind. . . . Now, ladies and gentlemen, we add sight to sound!"[67]

The age of commercial television had begun in earnest. Like Thomas Edison in the past and Steve Jobs in the future, David Sarnoff had built a complete ecosystem to promote this new media form: creation, programming, transmission, display. Farnsworth may have received the official credit as the inventor of television, but Sarnoff won the glory.

Educational Television

Starting in the 1950s, America saw rapid growth in the purchase of television sets. There were only a few thousand sets when they were first introduced, but just a decade later Americans had purchased well over 50 million of them. By the end of World War II, there was growing interest in using the public airwaves for educational purposes. In April 1952, the Federal Communications Commission (FCC) took action, allocating 242 television channels dedicated to educational programming, and commercial broadcasting stations began to emerge. Most were associated with universities or communities and offered limited programming.

Like films before them, television programs fell into three basic categories: entertainment programs designed for enjoyment, educational shows that raised awareness, and instructional programs that sought to teach a specific curriculum. Instructional shows were arranged in a series that built on each other and were planned with educators, with accompanying textual material and assessment.[68] This pedagogical model continues to hold to this day, even with material delivered over the Internet.

The US government and the Ford Foundation's Fund for the Enhancement of Education each spent over $100 million for classroom trials of educational television in more than 250 school systems and 50 colleges throughout the United States, involving over 300,000 students in the early 1960s.[69] The academic results of these trials were mildly encouraging, showing relatively small increases over traditional classroom teaching, but the main reason for wanting to instruct by television was its potential to spread one-time production costs over a larger audience and to take education from being a labor-intensive and expensive affair to one that could be manufactured and brought to scale. A Ford Foundation report calculated the break-even point, where televised lessons became cost-effective over traditional classroom instruction, was a mere 200–220 students, suggesting educational television to be an efficient way to educate students.[70]

But a number of problems needed to be overcome in order to use television to deliver instruction. The classes had to be closely synchronized with the broadcasters, and each broadcast channel could air only one lesson at a time. One particularly inventive solution in 1961 involved the use of airplanes equipped with television transmitters that constantly hovered 23,000 feet over a number of central Indiana schools, beaming videotaped instruction to them at agreed-upon times.[71]

Educational television proffered a new definition of the phrase "team teaching." Instead of two teachers presenting in the classroom, team teaching paired the classroom with a studio teacher who came prepared with the lesson plans and taught for 20 to 30 minutes each day. As an added benefit, the classroom teacher could spend more one-on-one time with students. This idea is a forerunner of what is today called the flipped classroom,[72] where students watch Internet-based videos on their own time and come to class to interact with the teacher.

Experiment in American Samoa

American Samoa is not the most logical place to launch the most ambitious implementation of educational television ever conducted. But the sleepy island in the South Pacific, about 2,500 miles south of Hawaii, was host to a "bold experiment" in education during the 1960s and 1970s.[73] As daring as the experiment was, it was also a classic example of how *not* to introduce educational technology into the classroom.

President Kennedy appointed Hyrum Rex Lee as the governor of American Samoa in 1961, and Lee found an educational system in shambles. Students were taught in rural one-room schools by teachers who did not have any formal training in education. Not a single teacher had a teaching certificate, or even tested beyond fifth-grade levels. Lee came to the island with a positive predisposition toward educational television. His daughter had learned how to touch-type from an educational television program, and she promptly secured an office job. Impressed by the process and his daughter's success, Lee completed a conversational French language television course.[74] "Television, it seemed to me," Lee remarked, "might be a way to bring about a quick upgrading of the educational system of our islands and an upgrading of the Samoan teachers at the same time."[75]

Using congressional funding, he built two studios that cranked out an impressive 200 lessons per week, each running 8 to 25 minutes long. The lessons ran the gamut from simple "talking heads"–style lecturing from a podium, to a smaller number of field-based sessions. In general, the students received approximately 8 hours of instruction by television per week, which worked out to about 30 percent of their total school time (fig. 1.24).[76]

Things began to unravel as the rigidity of the pace overran the students, with one teacher complaining that "the TV kept coming."[77] The teachers wanted more flexibility and autonomy, and they resented the centralized lesson planning, which resembled direct instruction pedagogy.*

* Direct instruction prescribes explicit steps for a teacher to follow when teaching a lesson.

Figure 1.24. Students taking a televised lesson in American Samoa. Courtesy of www.pagopago.com

The central office rejected the teachers' attempts to contribute to the planning, so teachers often divided their classrooms into smaller sections, using bookcases to partition the room for simultaneous television viewing, discussion, and supervised practice.

In the end, the adventure in American Samoa probably wasn't the "utter failure"[78] that Lee's successor, John Haydon, accused it of being in 1970, but it would be hard to call it a success. Education studies on Samoa showed that people can learn from educational television more or less as well as traditional classroom teaching. But some problems inherent in the medium need to be examined, in particular the lock-step pace that rigidly forces students to learn material at the same rate. The top–down, autocratic nature of the American Samoan experiment is typical of how many educational technology projects are implemented. Not only was there no initial consultation with the Samoan teachers in terms of curricular content and pedagogy, but they were also continually rebuffed in their attempts to provide feedback to the project leaders. That squashed any initial enthusiasm they may have had for the project.

Educational Broadcasting in the United States

Television's impact on education stateside was much less tied to school-based efforts than it was in American Samoa. Outside of the few efforts funded by the Ford Foundation in the 1960s, most educational television programming consisted of broadcast shows on allocated public television channels, many aimed at children, and they tended to be more entertaining than educational. A father's story, told at a Manhattan dinner party in 1966, would ultimately launch one of the most successful educational broadcast programs in history: *Sesame Street*.

An experimental psychologist named Lloyd Morrisett told his fellow dinner guests about something he had seen his daughter do some months earlier. Three-year-old Sarah had woken up early one morning, gone down to the living room, and turned on the family's television set. The day's broadcasting had not begun yet, and the classic test-pattern image of an American Indian chief flanked by graphics was frozen on the screen. When Morrisett found his daughter transfixed by the screen's image, he began thinking about television and children. At the party, Morrisett asked his dinner companions, largely made up of television producers, "Do you think television can be used to teach young children?"[79]

Morrisett's interest in television was not merely out of curiosity or parental concern. As a vice president at the Carnegie Foundation, he had recently awarded nearly $1 million in grants aimed at improving the education of preschoolers, particularly disadvantaged and minority children, and was growing frustrated about the inability to reach larger numbers of kids in spite of the amount of money they were spending.

The host of the Gramercy Park party was Joan Ganz Cooney (fig. 1.25), a successful producer of adult programs for WNET, New York's public television Channel 13. She was familiar with the problems involved in preschool education through her work on a documentary about the Head Start Program,* and Cooney and Morrisett quickly became coconspirators to help answer Morrisett's intriguing question.[80]

Cooney had studied education in Arizona, but aside from the short supervised training as part of her education degree, she had never taught in the classroom by herself. At age 25, she moved to New York City; through

* The Head Start Program is a US government program that provides preschool education for low-income children.

Figure 1.25. Joan Ganz Cooney with Big Bird and friends. Courtesy of the Sesame Workshop

her socially connected roommate, who "brought home directors the way ordinary people lug home groceries."[81] She was immersed in the heady world of 1950s Manhattan's elite, replete with weekends in the Hamptons. She attended one Long Island party at the mansion that F. Scott Fitzgerald used as a model for Jay Gatsby's palace in his novel *The Great Gatsby*. There she had a chance introduction that would change her career, saying later, "My path was always marked by encounters with strong men."[82]

That man was television mogul David Sarnoff. He had been successful in getting black-and-white television firmly entrenched in American society, and at the gentle prodding of the party's host, a close friend of Cooney's roommate, he asked Cooney to help him promote the introduction of color TV at the princely sum of $65 per week. Much as Sarnoff had done earlier working for Marconi's radio company, Cooney rose quickly at NBC Television, a later spin-off of RCA. Over the next 10 years, she moved from writing press releases and soap opera summaries to making more substantial contributions. Cooney knew the glass ceiling would prevent her from being an executive for any of the big three commercial television networks, so she moved to WNET, the New York City affiliate of the Public Broadcasting Service (PBS) in 1962 and began a successful career as a producer of socially important documentaries.[83]

A few days after that 1966 dinner party, Cooney and Morrisett began to plot a strategy to explore education television for preschoolers. Morrisett arranged with the Carnegie Foundation to pay Cooney $15,000 to write a proposal, and she took a four-month leave of absence from Channel 13 to travel to schools and universities, and to speak with television leaders. In March 1967, she submitted the 55-page proposal, titled "The Potential Uses of Television in Preschool Education," to WNET management, where it ultimately fell on deaf ears. The following month, with support from the Carnegie Foundation, she quit her job to find funding for the project. In the end she raised $8 million dollars, $4 million of it from the US government, and the rest primarily from the Carnegie and Ford Foundations.[84]

Cooney assembled an impressive team of advisors, including the cognitive psychologist Jerome Bruner, early reading literary researcher Jeanne Chall, as well as preschool teachers, community leaders, and television executives. They also made formative research an integral part of their production process, known as the Children's Television Workshop (CTW) / Sesame Workshop model, where the producers, educational content developers, and researchers work closely with one another in an iterative fashion.[85]

PBS debuted *Sesame Street* on November 10, 1969. It was a well-crafted mix of the three basic media genres: entertainment, education, and instruction. The production values rivaled any commercial television programming, with high-profile guests, lively music, compelling animated graphics, and, of course, the Muppets. There were "trips" to interesting places; informal moral guidance; and directed instruction on numbers, letters, logical concepts, and reasoning skills. For the first time, "education television was something children (and some parents) wanted to watch."[86]

Initial research studies to test *Sesame Street*'s effectiveness were positive, but soon fell into the typically murky waters of educational research. The venerable Educational Testing Service (ETS) evaluated the program's first year and reported generally positive results, cementing *Sesame Street* as an effective educational asset in the minds of most Americans. A subsequent ETS study,[87] and over 1,000 other studies conducted over the past 40 years, have cast some doubt on the program's true efficacy, but *Sesame Street* remains the most successful example of educational programming to date.

A recent study of children conducted by economists from Wellesley College and the University of Maryland found a high correlation between exposure to *Sesame Street* and better academic performance, making the bold claim that watching the show is as beneficial as attending preschool. But their positive results must be tempered by two factors, although one has to admire their cleverness in trying to get that kind of data, which is notoriously difficult to acquire. First, the way they determined whether kids had been exposed to the show was by its broadcast availability in their homes. One has to wonder about confounding factors here, such as whether areas without a PBS station may be poorer than ones with. Second, the researchers measured academic performance with data from the US Census by comparing students' ages with their grade levels in school, assuming that the less successful students are held back more often.[88]

One of the more ingenious introductions of broadcast television in the classroom is the Channel One project. This project intended to provide classrooms with well-produced current affairs programming. Media entrepreneur Chris Whittle, who had been concerned to learn that kids didn't know the difference between "Cher and Chernobyl," started Channel One in 1989, making a devil's bargain with school systems. In exchange for showing his advertisement-laden 12-minute cable-televised shows, Whittle would provide cash-strapped classrooms with television and video equipment. Channel One's offer opened a storm of controversy. Advertisers liked it because the audience size was 50 times the number of teens who watched MTV. There wasn't much academic content in the 12 minutes: more than half of the time was spent on ads for Coke, Twinkies, and Reeboks, as well as on promos and other filler. Left over was a scant 5 minutes for the actual educational content. Teachers were forced to play the shows at scheduled times, and they resented the complete lack of control over the content brought into their classrooms.[89]

Making Sense of Traditional Media

Beginning with that simple observation by Peter Mark Roget from his basement window in 1824, to Philo T. Farnsworth's sketch of the image dissector on the blackboard for his chemistry teacher a century later, one hopes it is clear how the technology behind moving images has evolved over time. The basic premise remains the same: movies and television are made up of a series of still images, that, when presented quickly, simulate the movement we perceive in the real world.

The steady transition from the optical-mechanical devices of Eadweard Muybridge and Thomas Edison toward the electronic devices of Philo Farnsworth and David Sarnoff has been one of increased abstraction—from images that were first captured as complete entities and recorded photographically, to abstract electrical representations that bore no physical resemblance to the scenes they sought to represent. This transformation allowed images to be stored and transmitted at great distances over the ether by the use of radio waves. This would ultimately pave the way for the next stage of abstraction—digitization—which completely removed the need for physical colocation and enabled imagery to be sent instantly anywhere on the globe via the Internet.

Educators have tried to use these new media forms at each step along the continuum to achieve better efficiency for instruction. The economic premise of using media in this way has a strong appeal, even to this day. Instead of re-creating the same lesson each time it is needed, it is created only once at a fixed cost, and it is replayed many times at a low marginal cost per student. This presumes, of course, that the same content is appropriate for all learners in terms of subject matter, scope, and difficulty—one of the main reasons these efforts have not been as successful as their promoters had hoped.

But the inherent constraints of the media's delivery vehicle may limit even the best-designed instruction. Traditional media's biggest limitation was that the learner had no real control over the media presented. At best, they could start, stop, and replay sections; if that media was broadcast, even a small amount of agency is lost. Educators began to look for new technology to overcome some of the restrictions that linearity imposed. Chapter 2 explores the promise of adding interactivity into the mix.

2

Interactive Media

It don't mean a thing, if it ain't got that swing.

Duke Ellington, 1931

The traditional media used in education has at least one major Achilles heel. No matter how well produced and instructional the content may be, the learning experience is primarily passive, and the extent to which learners can control the flow of that content is limited. For the broadcast media of radio and television, the situation is much worse: the only opportunity for viewer agency is to turn it on or off. When the content is recorded and made locally accessible, a viewer can replay segments of the instruction as many times as needed, and that control significantly increases, but overall the instructional experience is not optimal. This is because the media is inherently linear, designed to be consumed in a straight line from beginning to end.

This serious limitation of traditional linear media weighs heavily on instructors looking to create a more effective learning experience. A novel or movie may depend on this linearity to convey the narrative arc, which is essential to its form, but this can present an impediment to more instructional kinds of media. One way to overcome such limitations is to increase the learner's agency by making the media *interactive*. In recent times, that word has unfortunately become more of a marketing term than a descriptive one. For our purposes here, interactive media provides at least one key element that encourages rich user involvement: random access, or the ability to select any unique point in the media, and to be able to play or replay the media from that point.

A media system is also considered interactive when the learner is engaged in a form of relationship with the content, requiring "at least the appearance of two-way communication"[1] rather than a one-directional classroom lecture. Because most implementations do not allow for natural language conversations with the learner, the range of possible responses is limited, but what makes a system interactive is the user's ability to respond in some manner and in turn alter the presentation of content based on that feedback. In the best case, an interactive media system should be smart enough to estimate what the learner knows and accordingly adapt the instruction. Even the simple ability to let the user choose content parts from a menu can be empowering.

The Book as an Interactive Device

Comparing a book with a movie or a video is a useful exercise to explore interactivity. Even though both are best consumed in a linear, start-to-finish manner, the book contains some features that ease navigation. It contains words, grouped into paragraphs that can be scanned quickly for individual words, and it may have subheadings to further identity the ideas encrypted into those words. Each page has a unique number that makes it easy to identify and find. Those pages are organized into chapters, each one identified with a meaningful title, and the book often has a table of contents that directs a reader to the particular printed page where that chapter begins. At the end of the book, an index directs readers' queries to specific corresponding pages.

A film or video has none of these simple, but invaluable, navigational affordances. All we can do is start it, stop it, or scan it at a higher speed to find the parts we're looking for, but the interfaces for doing so are too crude, making the navigational process slow and cumbersome. In film and video, the unique identifier of a portion within the media (like a page number in a book) is replaced by a time from the start of the clip. Known as a *timecode*, that time is represented in a format of hours, minutes, and seconds, like this: 01:33:15. An interactive media system allows a viewer to play the media starting from any point, and stop at some other predetermined time. The time it takes to perform these actions varies depending on the technology used. Early videotape-based systems could sometimes take minutes to fast-forward or rewind to the desired spot, videodiscs could move to a new position in less than 10 seconds, and modern computer-based systems can instantly access any point within the media.

This ability to randomly access content offers a wide range of new capabilities to instructors, and many of them simply echo the navigational and search features that books already do quite well. Some systems provide access to the content by time (the page number), by searching for specific words (the index), and by identifying menu clips by a title (the table of contents). But the potential for transcendental change through interactivity comes when random access is paired with computer control.[2]

Computer control of media opens up the broad potential for instruction using media to be much more responsive, giving learners the option to choose sections to view, repeat, or ignore based on their individual needs and pacing preferences. Media controlled in this way makes it easier to implement the constructivist approaches advocated by modern learning theorists. Coupled with assessment, the ability to deliver specific media clips, at almost any granularity that random access affords, enables systems that can respond to learners' responses and provide customized content based on their needs.

Learning Sciences

Since the 1990s, a different view of how people might learn began to emerge from an eclectic group of researchers that included educators, psychologists, and cognitive scientists. They tried to tease out some of the internal workings of the learning process and to peer into the black box the behaviorists had so carefully constructed. These researchers were willing to speculate on some of the internal processes and to construct experiments to test those speculations, and a number of useful theories have emerged.

Cognitive Load Theory

Researchers have proposed that we have two primary pathways for sensing information: an auditory path, which takes information from the ears, and a visual path, which takes information from the eyes. The distinction gets fuzzy when reading is involved, because even though the eyes initially process words, the internal mechanisms of the auditory path are used to make sense of those words. Coupled with the fact that we have a limited number of things we can focus on at any given time, care needs to be taken to keep from overloading a single pathway with too much information.

These limitations in short-term memory have profound effects on how people are able to understand concepts, see patterns in data, and extract relationships between elements. To overcome these limitations, people have developed strategies that connect the limited short-term memory to the larger long-term memory by grouping together disparate items, called "chunking" (such as area codes, which classify telephone numbers by location), and immediately connecting new data to already known information (by linking to stories, schemas, and frames of reference, for example).

Cognitive load theory has emerged as one of the most important factors in understanding, learning, and paying attention overall. Research suggests that memory is composed of two primary structures—short term and long term—both of which are controlled by a central executive. Long-term memory is where instruction goes once it is actually learned by the user. Hence the aim of all instruction is to alter long-term memory, but information must first pass through the information gatekeeper, short-term memory.[3] Short-term memory is able to hold up to seven items (plus or minus two) at a time, and equally important is its ability to contrast, combine, or manipulate no more than two to four elements at a time. This is sometimes referred to as working memory. In addition, all the contents of working memory are lost within 20 seconds, without some sort of rehearsal.[4]

How People Learn

The National Research Council (NRC) funded an elite group of researchers to take a critical look at a wide range of learning research studies and apply their findings for practical use in the classroom. Headed by educational psychologist John Bransford, the group published their findings in an influential report, *How People Learn*,[5] that almost two decades later is still one of the best resources to explain complex learning processes. In the mid-1980s, Bransford had also done some pioneering work on learning using videodiscs, in *The Adventures of Jasper Woodbury*, which will be discussed later in the chapter. Like much of psychology, many of the report's findings seem obvious, but it is frightening how many of our traditional instructional settings flagrantly disregard its common-sense recommendations, and instructional media is no exception.

One of the report's key findings was that students come to class with preexisting knowledge and conceptions of how the world works, some of them wrong. Good instruction needs to meet students where they are, and build upon those preexisting understandings and knowledge.[6] Using the instructional media we have explored thus far, all learners are exposed to the same content regardless of the knowledge they bring with them. Traditional instructional media has to constantly walk a fine line between boring more advanced learners and frustrating those who come with less mastery of the topic at hand. The report also found that people learn best when actively constructing their own knowledge, rather than passively consuming content, which makes interactive media's increase in user agency an asset.

From Physical to Electronic Imagery

Up to this point, moving visual media has been largely mechanical in nature. When electricity was used at all, it was to either illuminate projection through arc or incandescent lighting, or to drive the mechanism using electric motors. The captured representation of the image was very close to the reality it tried to replicate. One could look at a movie frame with the naked eye and see the small image; even the squiggly lines on a phonograph record had direct physical relationship with the sounds it recorded.

As time went on, the correlation between the recording and its original subject became more and more abstract. If you were to look at an MPEG-encoded* video file from today's digital media, you would not have any clue as to its content by looking at the recording without elaborate mathematical decoding. In order to be transmitted like radio, visual imagery would need to be represented by analog electronics, and without using a physical medium like film. The recording of images via television is an intermediate point in the evolution toward digital media, but before there was television, there was *Phonovision*.

In 1981, a British engineer named Donald F. McLean checked out from his library a vinyl audio LP album from the British Broadcasting Corporation (BBC) on the history of early British television efforts. One of the discs, titled *We Seem to Have Lost the Picture*, contained a track that sounded like a "swarm of bees" when played.[7] The disc contained encoded imagery

* The Motion Picture Expert Group (MPEG) develops standards for video compression, used to reduce the size of digitized video clips and to make transmission of video over the Internet possible.

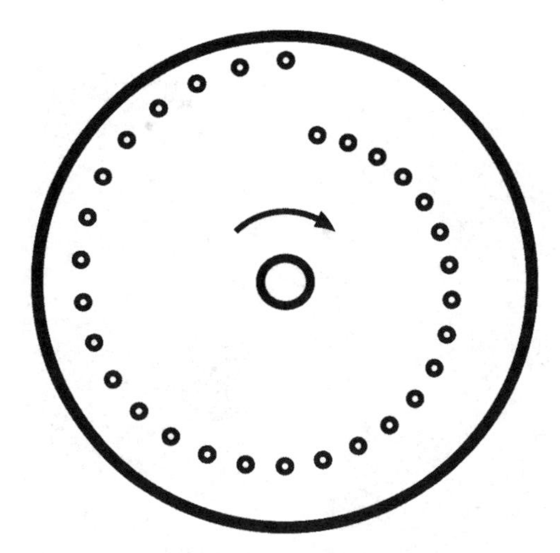

Figure 2.1. Illustration of a Nipkow disc. Courtesy of Majo Statt Senf

from a mid-1920s television system developed by the Scotsman John Logie Baird, who had succeeded in capturing images onto a modified phonograph disc but not in playing them back. This discovery launched McLean on a decade of experimentation to be able to see the imagery Baird had recorded in 1927 but never saw himself.

Rather than capture full images, as the earlier moviemakers did, Baird divided each image into small strips and recorded each one as a separate line. To achieve this scanning, he spun a large vertical disc that had 30 large lenses embedded within its circumference, called a Nipkow disc (fig. 2.1). The lenses were successively slightly closer to the center of the disc, so that each lens captured 1/30 of the image in a thin vertical stripe, which a photocell turned into electrical signals. Baird then used a modified phonograph to record 30 of the stripes as regular tracks on the record.

McLean methodically reconstructed Baird's images from the discs, and he was able to show the low-resolution moving pictures, which had never been seen before. One movie that Baird had recorded in 1927 was of Mabel Pounsford smiling for the "camera." Going one step further, McLean was able to find a contemporary photograph of Ms. Pounsford and compare it with the newly revealed imagery from 20 years earlier (fig. 2.2).[8] Baird's technique of scanning an image into discrete lines and turning those

Figure 2.2. Image of Mabel Pounsford in real life (*left*) and as recorded on Baird's Phonovision disc (*right*). Courtesy of Don McClean

lines into electrical signals would prove to be the basis of the better-looking images television would later offer.

Early Attempts at Random Access Media

Up through the 1960s, educational films and videos were typically 20 minutes long, with a number of scenes wrapped up within a larger topic. Educators knew random access of this rich media could be instructional, but they were stymied by the linearity inherent in the media itself. While it was possible to project only portions of a film in class, doing so was clumsy. Film projectors are difficult and time-consuming to thread, and they have no easy way to fast-forward to a specific scene. Videotape was slightly easier to set up, but it had the same search issues. Overall, neither was worth the effort for playing a short segment within a time-limited classroom period. The use of media was much the same from the individual student's perspective; students were unlikely to independently visit the library and view a portion of a film or video.

A number of solutions were offered to overcome the linearity inherent in using film and video for instruction, some bordering on Rube Goldberg–

Figure 2.3. An 8-mm film cartridge projector.

style solutions* to the problem of random access. A smaller gauge of film provided one possible solution. Called "the paperback of film" by early advocate John Flory, the smaller and shorter 8-mm film was one-third of the cost of 16-mm film. The smaller projectors were also much less expensive, but the big break for interactivity came when the film was loaded into plastic cartridges.[9]

The cartridges instantly plugged into a small desktop projector (fig. 2.3) that cost under $100 and was able to project onto a large screen or onto a smaller built-in screen that resembled a TV set. The films contained a soundtrack that was played through speakers or through a headset for private viewing. These projectors enabled so-called single-concept films to be quickly used in a classroom, or in library setting with headphones. Film producers created new titles or edited their larger educational and instructional films to fit into the new four-minute format. By 1968, over 7,000 titles were available, and many sold them for as little as $8 apiece.

* Rube Goldberg was a cartoonist known for his depictions of simple tasks done in convoluted ways.

Like many early media formats, such as videocassettes, 8-mm cartridges flourished for a while as the preferred format for hard-core pornography. But back at school, single-concept educational films never fully caught on beyond a flurry of popularity in the mid-1960s. The 8-mm projection companies failed to standardize the format of their cartridges, so that films packaged for one company's projector would not play on another's. The result was too small a market for any given format, and the critical mass needed for widespread adoption never surfaced. In addition, most of the schools' media budget was slanted toward buying larger 16-mm educational films that better supported whole-class instruction.[10]

In the late 1960s, a new way to achieve random access was tried using a centralized repository of audio players and, less often, video players. Each player contained a short segment of media that could be routed to the learner's viewing station by using remote, electrically connected terminals. A desired media player could be selected by pressing a dedicated push button, or by dialing up the player using an old-fashioned telephone dial. These *dial-access* systems were often used in language learning labs, allowing learners to select different instructional media segments on demand. When paired with printed workbooks, this system provided individualization of the media through adaptive branching to particular instructive segments based on assessments. Because each branch point required a dedicated player to deliver that media segment, these systems were expensive to implement, and the granularity of those segments was often too large to deliver truly individualized instruction.[11]

In the early 1980s, when video players became more affordable, there were attempts to control the videotape using a timecode. This provided random access to large amounts of media, but the time it took to find any given segment could take minutes, and the learner could experience long delays if the desired segments were on distant places on the tape. One particularly clever way to reduce that seek time, of which Rube Goldberg would have wholeheartedly approved, was to use multiple players, each one cued to the next position, with the device switching to the player that was cued nearest to the desired position.

Spinning Access: Videodiscs

Videotape, whether reel-to-reel or packaged inside a cassette, is recorded on lengths of plastic film up to 1,000 feet long. That physical reality made it time-consuming to find any desired point and provide practical random access to individualize the content within the reel because of the long times it took to fast-forward or rewind. To overcome this problem, developers looked back to an earlier physical arrangement of the content, and the flat audio disc that Thomas Edison had adopted for phonographs provided a good model. The content could be laid down in a circular pattern, and the player could move the "needle" to any point on the disc almost instantaneously.

Video information is significantly denser than audio, however, requiring more information to store the same span of time. While simply recording grooves on the disc could work in theory, only a few seconds of video could be recorded. Many inventions ensued in the 1970s that sought new ways to record information on the disc: tiny raised bumps, photographic mastering, engraved pits, and more sophisticated capacitance-based systems. The leading manufacturers of televisions—RCA, Sony, and Philips—all worked feverishly to develop disc-based players (fig. 2.4).

To be clear, the disc format was chosen primarily to deliver recorded video programs to consumers in a more convenient form than the video-

Figure 2.4. Magnavox videodisc player, circa 1980s. Courtesy of Marcin Wichary

cassettes popular at the time. It was only when customers abandoned the more expensive discs in favor of BetaMax and VHS videocassettes that the companies turned their marketing efforts toward training and education. Even so, in 1983, Sony's celebrated chairman Akio Morita boasted of the promise of interactive videodiscs for the masses, saying, "The viewer is no longer an observer, but rather an active participant who can shape the program being viewed."[12]

The drive and ambition embodied by David Sarnoff may have driven RCA's work on videodiscs at his eponymously named New Jersey research lab, and RCA produced a working videodisc in the mid-1970s. RCA faced fierce competition with the other manufacturers, in particular an alliance between Sony and Philips, which had a design that ultimately prevailed: lasers that read tiny pits from a reflective disc, the technique used by CDs and DVDs today.

All videodiscs shared some common elements. They were 12 inches in diameter, about the size of an old-fashioned vinyl record. They could play multiple tracks of audio and 54,000 video frames, which could be divided as 30 minutes of motion video, 54,000 still images, or any combination of the two. This ability to intermix still and motion sequences within the same medium offered many opportunities for instructional applications. Most importantly, the videodiscs could position and play from any point on the disc within a few seconds, allowing true random access of content in a practical manner for the first time.[13]

While analog videodiscs are no longer used in either the education or consumer arenas, they created a lot of enthusiasm for interactive media. The early educational videodisc systems tended to be expensive and clumsy, but the capabilities of random access, the mixing of still and moving images, and computer control provided a new basic interactive tool set. This led early adopters and researchers to explore how to effectively use these new capabilities for instructional purposes, and their findings formed much of the foundation that current interactive media is based upon.

Simulating Location with Media

In 1972, at the University of California, Berkeley, the urban designer Donald Appleyard teamed up with psychologist Kenneth Craik for a grant from the National Science Foundation. The two researchers wanted to simulate an urban environment by using still images shot while travers-

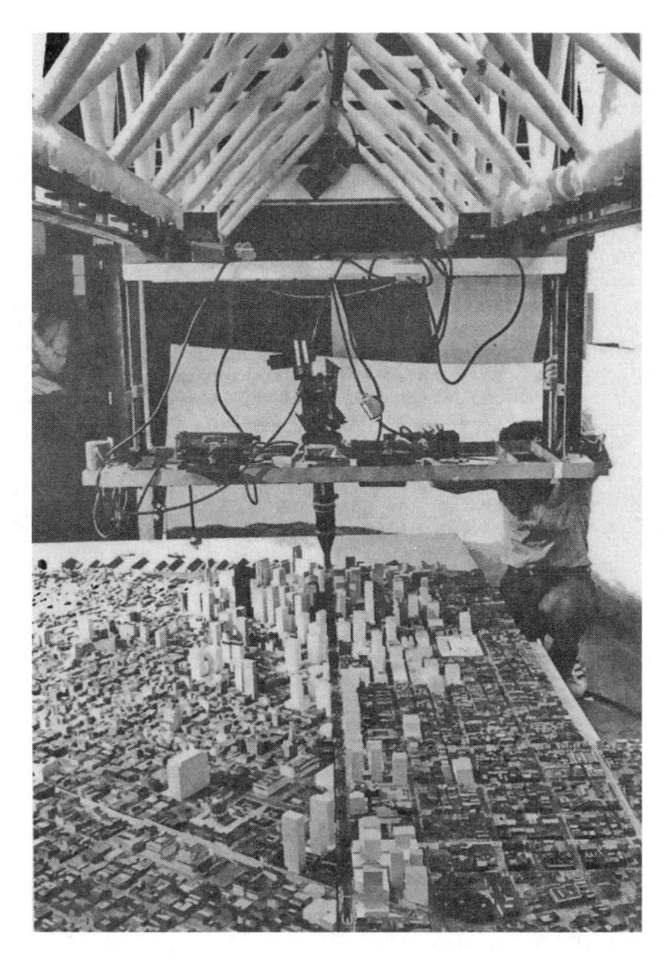

Figure 2.5. Modelscope camera filming a model of San Francisco. Courtesy of the Berkeley Environmental Simulation Laboratory

ing through a physical model. They developed a technique that used a special camera called a *Modelscope* that had a series of movable prisms and lenses that could be moved through a small, realistic model of an urban environment (fig. 2.5). These models were used to explore and elicit feedback and approval for unbuilt areas. As the camera traveled, it captured an image that could be played back in real time, simulating the appearance of smooth movement within the model's microworld. One of the Berkeley Modelscope researchers, the now-legendary cinematographer John Dykstra, went on to use exactly this same technique to film the spacecraft models in

Figure 2.6. Nicholas Negroponte, 2012. Courtesy of Fermín Rodríguez

George Lucas's first three *Star Wars* films, which to this day hold up well to projected scrutiny.[14]

As informative and immersive as the Modelscope films were, they were not very interactive and could show only the pathways that their creators chose to travel. If the viewer wanted to see what was to the right of an intersection but the film showed only the left, they were out of luck. Across the country, at the Massachusetts Institute of Technology (MIT), another group of researchers was looking at how to simulate a physical environment using a different technology: videodiscs. This work was one of the more innovative uses of videodiscs, simulating a sense of space and drawing upon the disc's ability to randomly access thousands of images shot from multiple perspectives to create a smooth path that a viewer could control.

One of the MIT researchers was Nicholas Negroponte, the son of a wealthy Greek shipping magnate who grew up in the tony Upper East Side of New York. He had attended MIT to study architecture (fig. 2.6) and soon became entranced with the emerging computer-aided design technology that helped architects design and digitally represent physical spaces on a two-dimensional computer screen. In 1968, he started a research lab at MIT called the Architecture Machine Group to explore just how people interacted with those systems. The lab attracted an eclectic group of brilliant researchers from a wide array of disciplines, such as music, computer science, architecture, and psychology. Later, the lab was combined with some other high-profile MIT projects, including Seymour Papert's Learning Program, Muriel Cooper's Visible Language Workshop, and Ricky Leacock's Cinema Verite Project. It was renamed the MIT Media Lab and blossomed into one of the premiere think tanks of the twentieth century for investigations at the junction of media and technology.[15]

A university research lab is an expensive enterprise to finance, and schools typically fund only a small portion of the costs. In the case of the MIT Media Lab, costs ran several millions of dollars a year, used to fund graduate students, postdoctoral fellows, professors, and expensive equipment. Labs typically use funding from foundations and grants from federal agencies to fill the massive gap, but the process of securing funds is long and tedious. In addition, such funding is sometimes an uneven revenue source for a sustained research effort. Negroponte found a new way to provide a steady stream of funding to support the Media Lab's expensive efforts: sponsored research.

Google is now famous for its 20 percent time work policy, where engineers are allowed to spend up to one day a week pursuing anything that interests them, with the company hoping that new product opportunities might emerge during this time of divergent thinking.[16] Decades earlier, MIT also had a 20 percent policy for their faculty members, but the fruits of this policy were not meant to line the school's pockets. MIT professors were encouraged to spend up to 20 percent of their time consulting on outside endeavors and actively participating in the creation of new companies. This arrangement allowed them to remain closely connected with what industry was doing and to jumpstart Boston's economic equivalent to Silicon Valley, the Route 128 corridor.

Negroponte raised the 20 percent concept to a new level, actively encouraging companies from all over the world to sponsor research at his lab. Companies such as Sony, Kodak, Xerox, RCA, and others were asked to fund specific projects. They could share in the discoveries found, but could not actively direct the research toward their own ends. The iconoclastic publisher Stewart Brand once described Negroponte as an "amphibian," a master at being equally comfortable in both academia and the boardroom.[17]

The amphibian was apparently equally comfortable with the military as he was with industry, and some of the work at the Media Lab was funded by the Defense Advanced Research Projects Agency (DARPA). DARPA had previously provided the funding and leadership for the pioneering efforts of academics and companies that had developed the foundations of the Internet.

On July 4, 1976, Israeli solders rescued over 100 passengers taken hostage at Entebbe International Airport in Uganda. To prepare for the raid, the soldiers had trained in a specially constructed full-size model of the airport, and the success of that mission piqued the military's interest in creating immersive simulations for training. Negroponte proposed a less costly training solution using images stored on videodiscs to provide a photorealistic sense of space. In the fall of 1977, he received a $300,000 grant from DARPA to test it.[18]

The MIT researchers used the same technique the Berkeley urban researchers had used on models, but in a move that preceded Google's Street View* by several decades, they outfitted a car with cameras as it traveled along real streets. The sleepy ski town of Aspen, Colorado, was chosen because one of the researchers, Bob Mohl, was from Colorado and thought the town was small enough (only 10 blocks by 15 blocks) to enable a complete tour of the town in a reasonable amount of time.[19] Four single-frame animation cameras were mounted at 90-degree angles to the car, and each shot a frame for every 10 feet of the car's movement (fig. 2.7). To minimize any discontinuities in lighting and distracting shadows, they filmed every day at noon, when the sun was directly overhead. Every street in the town was recorded twice, in the winter and in the summer.[20]

These images were burned onto videodiscs that were controlled by multiple touch-screen displays, using a computer that created an "interactive movie map" to navigate through the town of Aspen. Surrogate travelers could click on an overview map to instantly move to any point in the city, or leisurely "drive" down any street and look at the town from any

* Google Street View is an option on Google Maps that allows users to see locations as they appear from a street view. The images are captured by vehicles that drive methodically through neighborhoods and take 360-degree images with digital cameras.

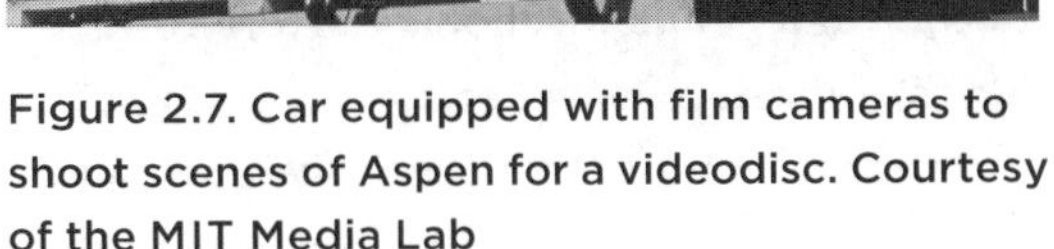

Figure 2.7. Car equipped with film cameras to shoot scenes of Aspen for a videodisc. Courtesy of the MIT Media Lab

Figure 2.8. Bob Mohl navigating the Aspen videodisc.

angle, making turns at any intersection (fig. 2.8). A button allowed the viewer to switch the season from summer to winter, and computer graphics provided a wireframe view on a separate monitor for context.[21] The result was an immersive experience that simulated actually driving through the town of Aspen. Unfortunately, not everyone thought the government's $300,000 was well spent. In 1980, the Aspen Movie Map project was a runner-up in Wisconsin Senator William Proxmire's famous Golden Fleece Awards for excessive government spending, presuming "there were generals frolicking with snow bunnies in Colorado."[22]

The late senator was not the only person less enthralled with Negroponte and his work at the Media Lab. Hypertext pioneer Ted Nelson once called it the "negentropy of Negroponte," because Negroponte had given the impression that his group had invented everything about interactive technology, mainly because of their penchant for constant demonstrations. Despite the sour grapes sound of it, Nelson did concede that the Media Lab and the Architecture Machine Group that preceded it "had been a fountainhead of research and design in the field of interactive systems. And, of course, they give good demo."[23]

The First Instructional Videodisc

At age 45, Dusty Heuston was a bit old to embark on the technological equivalent of an aboriginal walkabout. He was headmaster of the Spence School, an elite private school for girls on the Upper East Side of Manhattan, and had traveled a conventional career path: PhD in American literature from New York University, followed by a decade of college teaching at Brigham Young University (BYU), Vassar College, and Pine Manor College. But his walkabout in the summer of 1974 opened his eyes to the potential that the emerging microcomputer revolution could have on education and connected him with colleagues who would together create the first educational videodisc.[24]

Heuston was no stranger to technology; he had been using minicomputers at Spence since the early 1970s and was instrumental in setting up a network between Spence and six other independent schools in New York. He applied to the Sloan Foundation for a grant to travel across the country and explore how other schools, universities, and corporations were using technology in education. The application was a tough sell, as foundations typically fund large-scale research universities, not curious headmasters from elite private schools. Nevertheless, they offered him $14,000 to visit universities, large school districts, various experts, and executives in high-tech industry, including Intel's legendary founder Andy Grove. At Intel, he became a convert to the emerging potential of the microprocessor for education.[25]

At BYU, Heuston met with instructional psychology professor C. Victor Bunderson about computer-guided education in his office beneath the football stadium. After waiting nearly two hours, while Bunderson unraveled some thorny university bureaucratic issues, the two formed a close working partnership (fig. 2.9). Bunderson had long been interested in using technology in education and had helped develop one of the earliest computer-based tutoring systems, the Time-Shared Interactive Computer-Controlled Information Television system, or TICCIT (pronounced "ticket").

Before coming to BYU in 1972, Bunderson had been the director of the Computer-Assisted Instruction Laboratory at the University of Texas, and he used an IBM computer developed especially for instruction that would shape his thinking about integrating external image resources with computer-based instruction. The IBM 1500 Instructional System had a built-in optical slide projector that could present content that was rich in

Figure 2.9. Early videodisc author C. Victor Bunderson.

color, but static, side by side with dynamically generated computer graphic information on the cathode ray tube (CRT) display. The fact that slides could be randomly selected made it easy to create much more complex and visually rich instructional content that at the time could be done by a computer alone. This inspired him to look at videodiscs as an upgrade to that role.[26]

By 1977, the TICCIT project was winding down, and Bunderson was tiring of the overbearing university bureaucracy. On June 1, Dusty Heuston left the Spence School and moved to Provo, Utah, to join Bunderson to start a nonprofit company, WICAT, to develop educational technology products. Over the next four years they raised hundreds of thousands of dollars in grants from foundations, corporations, and the government to develop technology-based training systems using microcomputers. The personal computers of that time were underpowered for the task, so they built their own computer systems powered by microprocessors.

At an event in Manhattan one night, Heuston was talking with an executive from the Dutch technology manufacturer Philips about a new product they were working on, a videodisc player that could deliver thousands of still frames as well as full-motion video, all randomly accessible. The Philips executive offered WICAT two prototype players, $25,000, and a production assistant to create an educational disc that showcased the upcoming player.[27]

While working on TICCIT, Bunderson had developed a program for teaching college-level biology using programmed instruction that he wanted to convert to videodisc. In a programmed instruction lesson, the content is broken up into small parts, and the learner is led through them based on responses to embedded assessments. The videodiscs that Philips wanted to highlight had to play without being connected to a computer and had relatively simple interactive capabilities, lacking the ability for the kind of intelligent branching a computer might have.

The WICAT team drew upon the skills learners already had navigating books to help them learn to use the disc. Instead of being passively led through the content, the learners would need to actively seek the parts they wanted to view. A table of contents was printed on the videodisc's dust jacket, with disc locations that were easily entered on the player to allow users to quickly locate content. Once at a topic, the user paged to a question posed on the screen, which presented a series of keypress options as possible answers. The videodisc player could then branch forward to different frames for correct or incorrect answers, jumping further ahead or giving more help.[28]

WICAT partnered with publisher McGraw-Hill, which had previously produced a 16-mm film filled with compelling imagery that could be repurposed for the project. In the end, the disc cost four times more to produce than the $30,000 Philips had originally paid.[29] *The Development of Living Things* made its debut in May 1977, containing hundreds of still images and full-motion sequences, and it was successfully marketed by McGraw-Hill.[30]

Bunderson was later able to create a more robust version of the biology disc with the help of a grant from the National Science Foundation. He used a computer rather than simply printed words to navigate the disc, and he repurposed many of the techniques he had pioneered at the Texas lab and at TICCIT. These techniques included branching, interpreting natural language learner responses, and providing helpful status and feedback to the student. WICAT later conducted a study comparing students taught in a classroom setting with those who used the computer-controlled videodisc with the same content. They found that students who used the disc scored higher and had better retention than those taught traditionally, and they spent 32 percent less time learning the material.[31]

Videodiscs continued to capture educators' attention throughout the early 1980s, and two basic genres of educational discs emerged that mimicked the evolution of WICAT's biology discs. Level 1 videodiscs used inexpensive consumer-grade players, and they relied on printed materials as the primary means of navigation, whereas Level 3 discs required more expensive commercial-grade players that could be controlled by an external computer and offered much greater interactivity. Level 2—a videodisc player with a built-in microprocessor—was seldom implemented.[32]

The Jasper Woodbury Project

In the late 1980s, the Cognition and Technology Group at Vanderbilt University, led by the psychologist John Bransford and educator Ted Hasselbring, developed a series of educational interactive videodiscs, software, and supporting material called *The Adventures of Jasper Woodbury.* It was perhaps the most well-funded and carefully studied use of videodiscs in elementary education. It was one of the more successful projects using educational technology for the classroom because it was based on a solid theoretical foundation, involving classroom teachers in the design and implementation.

In its heyday, the group raised over $10 million, and with a staff of 80 people[33] actively investigated issues in learning and assessment of middle school math students, exploring alternative ways to accomplish both issues together. In particular, they focused on the problems in solving mathematical word problems, such as "Two trains leave different cities heading toward each other at different speeds. When do they meet?"

Word problems have always been the bane of middle school students, struggling or otherwise. In 1988, Swiss educational psychologist Kurt Reusser conducted a study where he gave the following problem to a group of middle school students: *There are 26 sheep and 10 goats on a ship. How old is the captain?* Even though not enough information was provided to correctly answer the question, over three-quarters of the students gave a numeric response. Working through a typical word problem requires a number of skills: identifying the facts needed, ignoring extraneous information (distracters), and coming up with a strategy—or set of smaller strategies—needed to solve the overall question.[34] There was a burst of activity in applying emergent personal computer technology to education, and the Vanderbilt group wanted to branch out beyond electronic versions of paper worksheets. They began by developing math simulation software, but soon found that the computers installed in schools (typically Apple IIs) were underpowered to deliver effective instruction. So they looked for a better solution.

A researcher at IBM's Watson Research Center introduced the Vanderbilt group to what IBM was doing with interactive videodiscs and the prospect of being able to deliver high-quality video in a random-access fashion, and they thought it might provide a way to deliver effective simulations to students. Using a disc containing the engaging story from the classic adventure film *Raiders of the Lost Ark*,[35] they provided a contextual backdrop and began to investigate how students generated mental models to solve mathematical problems that resembled traditional word problems. They used the freeze-frame capabilities of the discs to allow direct measurement of the screen, and were encouraged by the learners' progress in generalizing the solution found in one situated context to a different situation.

But as well produced and engaging as *The Raiders of the Lost Ark* was, it was designed as a piece of entertainment, not a tool for instruction, and copyright restrictions would have made any wider use expensive at best. In fact, the researchers wrote the film's director, Steven Spielberg, to ask whether he would collaborate with them on developing scenes to help teach kids math. They never heard back from the famed director, but they did get a terse "cease and desist" letter from his attorneys, asking them to immediately stop using the film in this way.[36]

The researchers then decided to produce their own videodisc to better serve their educational purposes. The result, *The River Adventure* used the same approach of using video to contextualize a problem within a real-world situation, and they began to work closely with classroom teachers to broaden its impact beyond the "well-defined word problem" and to provide an environment for investigating less-structured inquiry.[37] Encouraged by the results of *The River Adventure,* and responding to teachers' criticisms that the videos were too boring, the Vanderbilt group went Hollywood, embarking on "a series of successive approximations towards success" and "an ever-expanding lesson in humility."[38] Bransford and his colleagues received over $3 million in funding from the James McDonnell Foundation, the Eisenhower State Grant Program, and the National Science Foundation to produce a series they titled *The Adventures of Jasper Woodbury*. Over time, the group produced a dozen Jasper adventures, each one lasting about 15 minutes.

Teacher involvement was an important element in the development of the project. Media professionals actually produced the discs, classroom teachers reviewed the scripts for effective instructional approaches, and they even appeared in some of the adventures. This was important because effective use of the series required multiple class periods to complete, and teacher buy-in would be critical to its success. Teachers liked the authentic and connected ways in which the videos provided context to integrate math concepts, with one teacher commenting, "In life, things we learn are not separated. Jasper helps learning to be more like real life."[39]

Bransford and his colleagues defined their overall instructional approach as *anchored instruction*, which situates learning in an engaging, problem-rich environment that supports sustained exploration by students and teachers alike. Concepts are to be learned as tools, not facts to be memorized, and they explore why, when, and how to apply these tools to a diverse set of circumstances. The anchors offer a "macro context" that provides a common experience for people from different disciplines to talk with one another. A physics teacher might use the anchor to discuss issues from a scientific perspective, and a history teacher likewise can use the same anchor in a different context to discuss historical issues.

The overall goal of anchored instruction is to encourage students to become independent thinkers rather than simply be able to perform some basic calculations, such as computing the area of a circle, or "improving test scores in mathematics." Assessment is used to define the goals of the instruction by having students seek out subproblems, such as a circle's area in the service of solving the larger, overarching problem. This is done with scaffolding and support from the teacher, and it requires students to take a more active role in their learning. Research strongly suggests that actively acquired information is retained longer than information passively received.[40]

The group designed the Jasper Woodbury project under a set of principles that informed each adventure's development. These principles operated more as a Gestalt than as a discrete set of individual guidelines, with the overall goal to help students transform "mere facts into powerful conceptual tools."[41]

1. The adventures have a *video-based* format that presents more complex and interconnected situations intended to elicit a higher interest level from middle school students who are typically bored by simpler written word problems. Students liked the interactivity the videodisc provided, and the rapid seek times made it easy for them to find the data needed to solve the problems embedded directly within the stories.
2. The facts in the problem are *contextualized within a realistic situation*, rather than a dry rendition of the facts or presented in a lecture format. This makes the process engaging, and the skills learned are more likely to be generalized to other relevant situations.
3. The adventures encourage *generative learning*, the ability for the students to formulate appropriate subproblems on their own to solve the larger quandary. Generative learning promotes active versus passive learning techniques that challenge and build on students' earlier preconceptions, and it is often accomplished collaboratively in small groups.
4. All of the information needed to solve the problem is *embedded* within the adventure's video. The students are motivated by the story to seek out the embedded facts within and then solve the problem.
5. The problems themselves are *complex*, interconnected, and authentic to real-world situations. Each adventure has at least 14 steps or subproblems to solve the larger problem. Like modern video games, the steps provide intermediate success points to help kids' motivation and to let them learn to handle more complex situations.
6. The overall curriculum was *paired to standards* for middle school math education prescribed by the National Council of Teachers of Mathematics (NCTM).
7. The adventures are designed to involve *other disciplines beyond mathematics*. Because the narratives are richly situated in the real world, they encourage knowledge integration into history, literature, ecology, and the sciences.[42]

The videos are well produced, professionally shot and edited, and nicely acted, with a sense of folksy humor that is reminiscent of the 1960s *Green Acres* TV sitcom. The protagonist, Jasper Woodbury, is likewise a folksy rendition of Spielberg's Indiana Jones, with a ruggedly handsome but ap-

Figure 2.10. Frames from the "Rescue at Boon's Meadow" videodisc.

proachable demeanor, replete with a leather bomber's jacket but without the whip. The video transforms a traditional math word problem, albeit more complex than most, into a compelling story. The video provides the context and offers the data necessary to solve it. It also gives information that is not needed as a distracter. The learner is then asked to solve a problem based on that scenario and facts. "They're made like good detective stories," said Bransford, echoing some of the narrative genres used in education films of the 1950s. "You get to the end, and you realize you've already had all the clues. Students have to go back and search."[43]

The "Rescue at Boon's Meadow" adventure is a good example (fig. 2.10). The first scene shows Larry teaching his friend Emily how to fly an ultralight airplane in an open field, and he uses it as an opportunity to teach her the concepts of aerodynamics. This skill is not required to solve the problem, but facts embedded within the story may prove useful: air speed, space required for landing, miles per gallon, and so on.

Another scene is set in a restaurant, where we first meet Jasper, who is celebrating Emily's first flight with Larry. Jasper mentions he's going to drive 60 miles to go fishing. The scene ends with the three friends divvying up the check, a distracter to the ultimate problem. While fishing, Jasper hears a shot and finds a wounded bald eagle, and he calls for help on his CB radio (this was before the days of cell phones), knowing that time is of the essence, saying, "Call Emily Johnson, she'll think of something."[44] Interestingly, the bald eagle used in the scene's filming had been rescued from a similar real-world hunting accident.[45] We see Emily mulling over the facts in her mind, building the subproblems needed to solve the main one. A new set of fact exposition ensues, and the problem is finally unveiled: *What is the fastest way for Emily to retrieve the injured bird?* The check-splitting scene, a distracter for the larger problem, provides an opportunity for using the same expensive-to-produce video for other, less involved, problem-solving activities.[46]

The adventures can be used in whole-classroom instruction using traditional linear videos or DVDs, but the rapid search and freeze-frame capabilities of the videodisc make the story a more readily available source of the raw data needed for solving the problem. Students may use a number of tools to search through the scenes on the videodisc: controlling the player via a remote control, using a scanner to read printed bar codes in an index, or operating a personal computer connected to the videodisc player. The group also developed a number of computer applications, most of them based on Apple's HyperCard software, to provide a series of SMART challenges (Special Multimedia Access Arenas for Refined Thinking) that used computer-generated graphics to extend Jasper's rich video scenes.[47]

As a side note, my research into the Jasper project pointed out just how ephemeral educational technology can be. As popular as the videodisc format was when the Vanderbilt group first produced the project, few people use them today. Plenty of the original Jasper videodiscs could be found, but I was unable to locate a player nearby that could play them. I was able to find a CD-ROM version and, with a little wizardry, was able to play the "Rescue at Boon's Meadow" disc, but none of the SMART software tools would operate anymore, on either a contemporary Mac or Windows PC.

This is a recurrent problem when using technology in areas that were once the domain of more stable media, such as printed books. Even the most ardent proponents of using technology recognize the short lifespan that digital products can have. University of Virginia English professor Jerry McGann's archive of the English poet and illustrator Dante Gabriel Rossetti's work was a groundbreaking digital project[48] that made Rossetti's work easily accessible for scholars to study. But McGann once commented to me that the only way he would be able to truly archive his project was to print out each page of the website and bind it for the library to store as a traditional printed book.

The Vanderbilt group studied the effects of using *The Adventures of Jasper Woodbury* on a wide group of students throughout the project's development. In educational research, assessment is a tricky issue. The goal is to determine what effect, if any, your intervention had on students, overcoming the tremendous variation in teaching techniques and in the students themselves. Because the interventions are short and sharply focused, it is difficult to see influences in a broad measurement, such as on a standardized test. For this reason, researchers often develop their own assessments that target the kinds of change they presume their intervention will produce. Unfortunately, these homegrown assessments do not often match broader standardized tests, and they tend to show bigger gains than the broader tests do.

In 1990, the Vanderbilt researchers launched an ambitious nine-state evaluation of the project to understand how teachers and students used Jasper, how they compared with students who didn't use it, and how the internal assessment they developed meshed with the different states' standardized tests. As expected, students who used the Jasper adventures did better than students who didn't, particularly on word problems. Students who used Jasper connected with a computer did much better than students who used Jasper alone. More importantly, students who used Jasper either way showed much less anxiety toward mathematics than students who didn't. The results that compared student performance on standardized tests showed the dreaded but predictable NSD, for "no significant difference."[49]

Ultimately, using *The Adventures of Jasper Woodbury* in a classroom is more about adapting the anchored instruction approach than about adopting a specific technology or media. The research its creators collected offered insight into the value of using visual media to contextualize complex problems and make them more accessible to students at all levels. The data embedded within the narrative mimic the nature of real-world problems, and the skills learned in solving those problems provide tools to employ in the real world and help strengthen the hard-to-acquire metacognitive skills required for complex problem solving. The situated nature of the data makes it easier to involve disciplines other than mathematics into the classroom, further encouraging the generalization of the skills.[50] The Jasper project is still available on CD-ROM,[51] some 30 years after they were first produced, and they hold up pretty well to modern scrutiny. There are plans in the near future to release the footage into the public domain for anyone to use.[52]

A number of other educational projects using videodiscs sprang up, as well. The largest teacher's union, the National Education Association (NEA), and one of the big three broadcasters, ABC, teamed up for an ambitious but ill-fated videodisc project called the *ABC/NEA Schooldisc*. They planned to create 20 one-hour magazine-style videodiscs to be released to schools every two weeks for a year. Each disc contained six 10-minute segments, each of which was dedicated to a different discipline—language arts, science, math, social studies, the arts, and current events—using footage from ABC's vast video archives.[53]

Most educational technology efforts suffer from a lack of teacher involvement, but at some point there can apparently be too much of a good thing. Because of the NEA's involvement in the project, the curriculum and overall content production were overseen by a small group of teachers, with too little involvement of the production professionals and instructional designers. The end result was an expensive, poorly produced product that was never embraced by the schools, and the project was quickly scuttled.[54]

The National Gallery of Art

Schools were not the only early adopters of videodiscs for education. Museum educator Ruth Perlin was intrigued by the possibilities of using the new technology when executives from Sony demoed their upcoming videodisc player at an event in 1973. She saw the disc's potential to instantly present thousands of high-quality images of artworks and video clips as benefiting art education, and she "waited and waited and waited" for the technology to appear in the market. A few years later, Perlin joined the National Gallery of Art in Washington, DC, as a curator for national education programs. In 1979, a media company from New York approached the gallery about issuing a videodisc that re-created the large volume *National Gallery of Art*, published a few years earlier by Harry N. Abrams Inc., that included some 1,600 works from the gallery's holdings.

Museums walk a fine line between wanting to encourage in-person visits and making their rich resources available remotely for teachers and students who are unable to visit. They want to serve their distant constituencies, but at the same time, many institutions don't want to discourage the in-person experience and are often wary about using technology to essentially give away their content. The National Gallery of Art was no exception. Perlin had a meeting with the museum's director, the aristocratic but populist director J. Carter Brown, who enthusiastically endorsed the idea by saying, "Do it!" but at the same time, he kept the project "under the radar." The company paid for the production costs, and Perlin's group

Figure 2.11. Voyager president Bob Stein demoing the NGA disc on the *Computer Chronicles* television show in 1987.

chose the works to highlight, digitized the artwork, and wrote biographical scripts on the artist's lives that the producers used to create short films.

The videodisc was published in 1983, highlighting thousands of individual works of American art, and by all accounts it was wildly successful. Most American schools at the time had videodisc players, and the gallery made 200 discs available to schools on long-term loans. As with *The Adventures of Jasper Woodbury*, teachers were encouraged to use the discs across the curriculum, not just in art education class.[55]

The gallery created a follow-up videodisc in 1993, using a grant from the Annenberg Foundation, and it was the first videodisc to use digital imaging as the basis for the image catalog. Encompassing over 2,600 works of art—all of the museum's American paintings, sculptures, and a vast selection of works of art on paper—it contained over 10,000 still frames, including full-frame images and close details highlighting brushstrokes, signatures, and other fine details. Perlin believed it was important to present the artists as real people, and she included video clips showing scenes of their houses, studios, and sometimes interviews with them (fig. 2.11).[56]

They made an additional 2,500 copies available to schools in a politically savvy manner. To celebrate their fiftieth anniversary, the National Gallery of Art gave copies of the discs to schools from all 50 states, with congressmen and senators personally choosing which schools would receive the discs. Perlin's team at the gallery subsequently developed a HyperCard companion for the videodisc—with artist biographies, keywords, and object information—to encourage its use across the curriculum.[57]

Making Sense of Interactive Media

Videodiscs remained popular in medical and training applications for a few years, but outside of a few noble efforts such as *The Adventures of Jasper Woodbury*, they never caught on in the larger educational community. As with the first educational films some 75 years earlier, there were not enough videodisc players in schools, and also a paucity of good-quality discs that would encourage schools to purchase them over the well-entrenched 16-mm films and projectors. Another issue was the change in teaching that interactive media required for effective use in the classroom. Noninteractive films and videos are passively viewed as a whole-class activity, often providing a break for the instructor. Interactive media is typically used individually, or in small groups, requiring a different set of teaching skills with which the instructor may not be as familiar.[58]

There were a number of studies that looked at the effectiveness of using videodiscs versus more traditional instructional techniques, and, like much research in educational technology, the results were mixed, with most showing no significant difference between the two. One study found a large gain across students of all ability levels, with larger gains for lower-ability students. The study cautioned, however, that the term "interactivity" was too often used to describe the hardware being used, rather than the actual method employed by the instructor and learner.[59]

The interactivity provided by the videodisc's ability to randomly access any scene within a few seconds made linear video more accessible and effective for educational uses. But when that interactivity was linked with a computer, a whole new kind of learning tool began to evolve. Chapter 3 examines the pairing of media with a particular kind of software to create a new genre of interactive media known as *hypermedia*.

3

Hypermedia

In Xanadu did Kublai Khan
a stately pleasure-dome decree,
where Alph, the sacred river, ran
through caverns measureless to man
down to a sunless sea . . .

Samuel Taylor Coleridge, 1816

The Media Lab's Nicholas Negroponte once remarked that "an optical disc without access to a computer is like an airplane without wings; it's not a meaningful machine."[1] Those wings took on a special shape in 1987, when Apple Computer introduced HyperCard. HyperCard was a graphic-authoring software application that allowed users to instantly show specific pages, or "cards," of information by clicking on links, releasing a world of information to be randomly accessed. Important to this discussion is that a link on a HyperCard could also remotely control external devices such as videodiscs, opening up a rich trove of media to be summoned instantly for the viewer. Today, we all take it for granted that clicking on an underlined word on a Web page will take us to a new site on the Internet almost instantly, but the concept of hyperlinking had a legacy dating to well before the Second World War.

As We May Think

Vannevar Bush (not a close relation to the Bush presidents) was born in 1890 in Everett, Massachusetts, and would become the most politically connected American scientist and inventor since Benjamin Franklin

Figure 3.1. Vannevar Bush, 1940s.

two centuries earlier. He was the son of a Universalist minister who was also an active Freemason, descended from a long line of New England seafarers. The younger Bush (fig. 3.1) seemed to fault his father only for the unusual name he had chosen for him, often wishing he had simply been called "John." He particularly hated its original Dutch pronunciation of "van-NEE-var" and was known as simply "Van" to his friends, joking that they were not able to pronounce his proper forename.[2]

Bush was a mathematically gifted child, and, like many of his time, he was smitten with the emerging radio technology. This interest probably influenced him to eventually study electrical engineering during his undergraduate studies at Tufts University. While still a student in 1912, he patented his first invention, a clever device called a "profile tracer,"[3] which could accurately draw the boundaries and elevation of an area of land as the wheelbarrow-like device was pushed about. This profile tracer would be the beginning of his interest in using mechanical methods to calculate mathematics. Unfortunately, despite his efforts to market the device to commercial companies, and to Bush's great disappointment, no one ever licensed it.

Bush entered the graduate program at the Massachusetts Institute of Technology in 1915 and within a year received his doctorate in electrical engineering. After graduation he went back to his alma mater, Tufts, to teach for a few years, but then returned to teach at MIT four years later. While there, he continued working on methods to solve mathematical problems by creating a series of mechanical calculators. These massive machines used gears and rotors to solve the tough differential equations needed to calculate the trajectories for battleship guns, which were time consuming and error prone to do by hand. In time, analog electronics took over from mechanical devices to perform the time-consuming calculations far faster than either mechanical or human "computers."

Even when digital computers emerged some years later, Bush was skeptical of them, favoring mechanics over digital electronics. In fact, during his work presiding over the Manhattan Project during World War II, he objected to funding the ENIAC project, the first true computer, because he thought the thousands of vacuum tubes it needed would be unreliable. His biographer, Zachary Pascal, wrote, "Analog devices had a powerful psychological hold on him. They linked him to a 19th-century world of gears and metal—where knowledge was a direct physical encounter of the head and hands."[4]

Bush may have been slow to embrace the digital computer, but he presciently saw technology as the answer to help organize the world's ever-growing mass of information. He was particularly captivated by a new form of photography known as microfilm, which could store huge numbers of documents in a very small space and also allow rapid access to them. In 1936, he sent a proposal to the Federal Bureau of Investigation to develop a system to store that agency's massive collection of fingerprints on electrically controlled microfilm viewers, but J. Edgar Hoover personally turned him down.[5]

In 1939, Bush left academia to take the helm of the prestigious Carnegie Institute in Washington, DC, and that proximity to national power launched a new trajectory toward becoming the most influential scientist in government. His $25,000 salary from Carnegie (around $300,000 today) allowed him to consult at no cost to the military and to the government as a "dollar-a-year man." He soon forged a close relationship with Franklin Roosevelt, proposing and then heading up the National Defense Research Committee to connect scientists with the military. There, he played a key role in convincing Roosevelt to start the Manhattan Project, and in 1941, he led the effort that resulted in the first atomic bomb.[6]

With World War II at an end, thanks in large part to his work on the Manhattan Project, Bush turned his attention away from wartime efforts and wondered what was next for him, musing: "This has not been a scientist's war; it has been a war in which all have had a part. The scientists, burying their old professional competition in the demand of a common cause, have shared greatly and learned much. It has been exhilarating to work in effective partnership. Now, for many, this appears to be approaching an end. What are the scientists to do next?"[7] For Bush, the next step was to explore people's relationship with the ever-growing amounts of information piling up around them.

In July 1945, Bush wrote an essay for the well-respected literary magazine the *Atlantic Monthly* that inspired and resonated with innovators for decades after it was first published. "As We May Think"[8] sounded the opening salvo of a new way of navigating the large bodies of knowledge people were generating. Bush believed that "knowledge evolves and endures throughout the life of a race rather than that of an individual"[9] and that knowledge was coming in at such a great pace that no individual could absorb, much less remember, much of it. He noted that although this ever-increasing degree of specialization was necessary for progress, it made any possible connections between disciplines superficial at best. And he believed that technology could help.

Drawing on earlier insights and ideas from his work with mechanical computers in the 1930s, Bush had offered a version of his article to *Fortune Magazine* in 1939 under the title "Mechanization and the Record," but he was not able to convince the editors to publish it, and he was later swept up in the war effort. At the end of the war, he again pursued publication with *Fortune*, but the magazine was unsure about its fit with their readers; Bush ultimately went with the *Atlantic*. The article was republished soon afterward by the popular *Life Magazine*, in a shorter, more accessible form, complete with drawings of what the proposed device might look like.[10]

As with the work of Philo Farnsworth and Vladimir Zworykin in early television, Bush recognized early that mechanical devices were no match for electrons. He envisioned a machine, which he named the Memex, that could store and instantly retrieve vast amounts of information, and was operated by a knowledge worker seated at a large desk (fig. 3.2). The goal was to use technology to amplify memory and extend human cognition. The user of the Memex could instantly browse through an almost unlimited

Figure 3.2. Full-size model of Vannevar Bush's Memex, made in 2014 by Trevor F. Smith and Sparks Web.

number of articles, personal notes, and images using two independent viewing screens, each the result of optically projected microfilm. A unique label identified each of the stored information items. The user could then connect an item on one screen to another, creating a permanent "link" between their labels. These links were preserved in "trails" that captured the search process and became another searchable item for further navigation.

Bush fashioned the Memex to replicate how he believed the human mind actually worked. Rather than indexing every item directly in a kind of table of contents, he proposed that the mind operates by establishing associative links between items in memory, and the brain is a collection of trails of these links, permitting "selection by association." Although the storage medium of his day was microfilm, which could not easily be recorded to, the Memex could add a new note or photograph by capturing a "dry photograph" image of the new item placed on a transparent screen, just as a present-day scanner does.[11]

His ideas of links and trails provided the fundamental inspiration for the kind of hypertext linking used to navigate sites on the Internet, explaining that "any item may be caused at will to select immediately and automatically another. This is the essential feature of the Memex. The process of tying two items together is the important thing . . . Thereafter, at any time, when one of these items is in view, the other can be instantly recalled merely by tapping a button below the corresponding code space. . . . It is exactly as though the physical items had been gathered together from widely separated sources and bound together to form a new book. It is more than this, for any item can be joined into numerous trails."[12]

It would take nearly half a century, but the Internet has fulfilled many of Bush's prophesies about fast, linked access to the world's information. Although Vannevar Bush never actually built any of the devices he described in his essay, many of his pioneering ideas can be linked directly to innovations today, including the Internet, image compression, and even Google Glass.* More importantly, "As We May Think" served as a profound inspiration for the next generation of information research, and especially instructional media.

The Mother of All Demos

Douglas C. Engelbart was one of many who was inspired by Vannevar Bush's forward-thinking essay. At the end of World War II, the 20-year-old Navy radar technician was awaiting transport back to the United States when he came upon that June issue of the *Atlantic Monthly*. Years later he wrote Bush, saying, "I might add that this article of yours has probably influenced me quite basically. I remember finding it and avidly reading it in a Red Cross library on the edge of the jungle on Leyte, one of the Phillipine [sic] Islands, in the fall of 1945. . . . I re-discovered your article about three years ago, and was rather startled to realize how much I had aligned my sights along the vector you had described. I wouldn't be surprised at all if the reading of this article sixteen and a half years ago hadn't had a real influence upon the course of my thoughts and actions."[13]

On his return from the Philippines, armed with his degree in electrical engineering and the then-cutting-edge knowledge about radar he had learned in the Navy, Engelbart worked for a predecessor of the National

* Google Glass was an experimental device introduced by Google in 2013, where a computer was embedded within eyeglass frames and afforded its wearer constant interaction. See https://en.wikipedia.org/wiki/Google_Glass.

Aeronautics and Space Administration (NASA) as an electrical engineer. His office was based at Moffett Field, a Navy airfield in the heart of what was then the world's largest prune orchard. It would later become known as Silicon Valley. Engelbart continued to ruminate about ideas on information management, and he developed a new personal goal. The hard task ahead was no longer the accumulation of new knowledge, but in knowing how to find the answers already stored somewhere. He began to sketch out what a system augmenting human cognition might look like, using computers to draw manipulatable symbols on the screen that could actualize some of the innovative ideas Bush had first described decades earlier.

Engelbart received his PhD in engineering at the University of California, Berkeley, where some of the first computers were being built. He didn't see himself as an academic and instead sought a commercial avenue to pursue tools that augmented human intellect. His opportunity arrived when he joined the Stanford Research Institute (SRI) in October 1957, the same month the Soviets launched the Sputnik satellite. SRI was then an institute within Stanford University that focused on research in a wide range of areas, from finding a substitute for natural rubber to economic studies for industry and the military. It was the perfect place for Engelbart to work out his ideas on using the emerging computer technology to augment human intellect.[14]

In 1963, he submitted a proposal to one of SRI's funders at the Air Force, outlining his ideas on how to do this from philosophical, theoretical, and pragmatic perspectives. Engelbart situated his proposal squarely within Bush's 1945 essay, saying, "In six and half pages crammed full of well-based speculations Bush proceeds to outline enough plausible artifact and methodology developments to a very convincing case for the augmentation of the individual intellectual worker."[15]

He described a conceptual framework with four principal ways that people use tools:

1. As *artifacts:* physical objects to be manipulated.
2. With *language*: attaching symbols to mental models of concepts and things in the world.
3. Using *methodology*: the strategies for organizing goal-centered activities.
4. Through *training*: using instruction to marshal the other three toward a productive outcome.

Engelbart proposed a computer-based system capable of automatically manipulating these interdependent symbols and concepts in a hierarchical structure. To make these abstract ideas more concrete, he employed a demonstration of the proposed device with a "friendly fellow" named Joe to help ground the discussion.[16]

The proposal resonated with the Advanced Research Projects Agency (ARPA), which in 1964 gave SRI access to one of the new time-share computer systems and $500,000 to fund the Augmentation Research Center. Dwight Eisenhower created DARPA in 1958, to fund cutting-edge research projects, and was directly responsible for funding the early development of the Internet. Engelbart's Augmentation Research Center was one of the first nodes on the early version of ARPANET, which ultimately evolved into the Internet we use today.[17]

For the next four years, Engelbart and the other 17 members of the Augmentation Research Center worked hard at creating the system he had envisioned nearly two decades earlier. On December 9, 1968, Engelbart showed the fruits of their labor at an elaborate demonstration to the Fall Joint Computer Conference in San Francisco, known later as "the mother of all demos." Engelbart stood alone on the stage in front of a computer screen, wearing mission-control style headphones to communicate remotely to team members back in Menlo Park via a microwave link connected to six networked computers (fig. 3.3).

The 90-minute demo was a tour de force of the presentation technology of the day. Thankfully it was filmed for posterity, showing a mouse-driven system capable of manipulating words and paragraphs. It was the first demonstration of a graphically oriented interface that allowed direct user control. We take this system for granted today in even the most basic word processors, but in 1968, the abstract concept of symbol manipulation was groundbreaking. Engelbart's presentation earned a standing ovation. The mouse-driven system personified computing, especially when compared with the slow punch-card interfaces then popular, and it ushered in the age of the personal computer.[18]

Engelbart left SRI for a division of McDonald Douglas, and the other members of his Augmentation Research Center went to work at Xerox's Palo Alto Research Center (PARC). In the 1970s, his former team developed the innovative Xerox Alto and Star office computers that refined the system he had first demonstrated in San Francisco.[19] In the 1980s, Apple

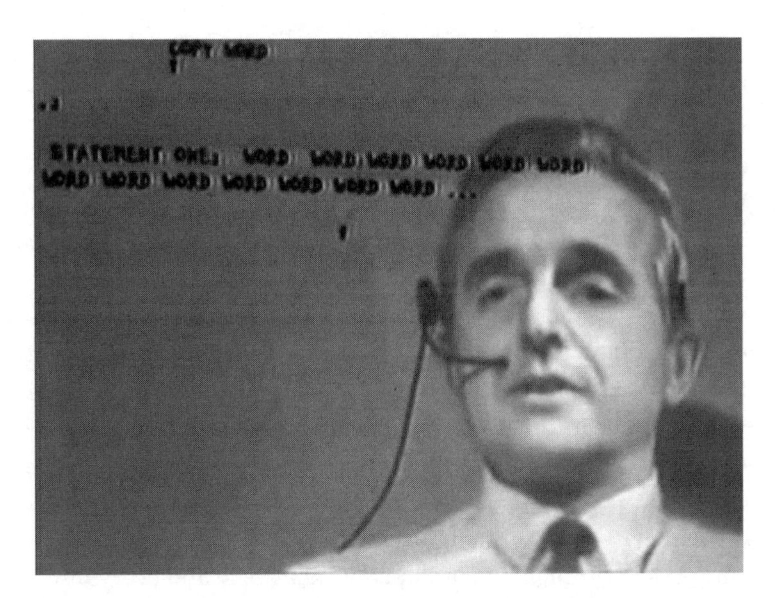

Figure 3.3. Douglas Engelbart giving "the mother of all demos," 1968.

Computer entered into an investment arrangement with PARC: in exchange for allowing Xerox to invest in the smaller, but rapidly growing, company, Apple could gain access to PARC's graphical user-centered technology for its Lisa and Macintosh systems, and the world of personal computing began to look much like it appears today. Douglas Engelbart was well known for his pioneering efforts within the technology world, but the greater public is less familiar with his contributions.[20]

Xanadu

Douglas Engelbart wasn't the only person profoundly inspired by Bush's "As We May Think" essay in the *Atlantic*. Eight-year-old Theodor Nelson first heard about the Memex from his grandparents at the dinner table, where they would often read aloud articles from magazines and newspapers. Nelson would later go on to design and implement his own vision of linked information, and would become known, along with Engelbart, as one of the founders of hypertext.[21]

But it was an earlier experience at the age of 4 that had a more profound impact on Nelson than Bush's article. As he sat in the back of a boat his grandfather was rowing, Nelson dragged his hand in the water behind, watching wondrously as the water current trails joined, separated, and rejoined, and "were no longer in the same way."[22] This became a transcendent experience for him, one he remembered for almost 70 years, and was the beginning of his profound thinking about the connections between things.[23] Nelson's remembrance is reminiscent of a lovely short story by the novelist Salman Rushdie, "Haroun and the Sea of Stories." In the tale, a young boy is traveling in a boat in the ocean, and a genie tells him, "The sea is full of a thousand different currents, the Streams of Story. These are 'all the stories that had ever been told and many that were still in the process of being invented.' "[24] The streams represent the countless subparts of a whole that can be infinitely reorganized to create new assemblages.

Nelson has described himself as a dreamy, lonely, and unathletic kid, raised by his elderly grandparents in the trendy Greenwich Village section of Manhattan. In spite of being bright, he claims to have hated school "from first grade through high school, unrelentingly and every minute."[25] His mother, the actress Celeste Holm, was a famous leading lady of the time, starring in *All about Eve* (1950) and winning an Academy Award for her role in *Gentleman's Agreement* (1947). His father, Ralph Nelson, worked in the film industry, directing the first adaptation of a major play for television, *Hamlet*, in 1959, and later *Requiem for a Heavyweight* (1962) and *Blue Soldier* (1970). Ted Nelson ruefully commented, "Many children fantasize that their real parents are faraway glamorous people. For me this was actually true. Like Harry Potter, I had magical parents that were not present, and like Harry Potter, I have been greatly punished for it."[26]

Nelson has been characterized charitably as having a highly active mind that easily darts between topics, the way millennials flit between digital screens (fig. 3.4). "There are no stops in the flow of his speed, only commas, dashes ellipses." Nelson has severe attention deficit disorder (ADD), and seems to be proud of that diagnosis, although not its name, saying it was coined by "regularity chauvinists," defining them as "people who insist that you have got to do the same thing every time, every day, which drives some of us nuts. Attention Deficit Disorder—we need a more positive term for that. Hummingbird mind, I should think."[27]

By all accounts and especially his own, Ted Nelson is not a modest man. He wrote in the forethought section of his 2010 autobiography, "Everybody wants to tell their story; I have special reasons. I have a unique place in history and I want to claim it. This is not a modest book. Modesty is for

Figure 3.4. Theodor Nelson, 2011. Courtesy of Gisle Hannemyr

those after the Nobel and that chance (if any) is long past. This is what I want known long after."[28]

In an influential paper given at a computer conference in 1965, Nelson coined the term *hypertext* to describe "a body of written or pictorial material interconnected in such a complex way that it could not conveniently be presented or represented on paper." He also coined the term *hypermedia*, applying that same nonlinearity to films, videos, and audio recordings that can be "arranged as non-linear systems—for instance, lattices—for editing purposes, or for display with different emphasis."[29]

An important part of his definition is the word "nonlinear," which he hoped could free ideas from the constrainment that representation on paper forced. Like Bush and Engelbart, Nelson believes that ideas are represented nonsequentially in our associatively structured minds. Hypertext and hypermedia are ways to free their depiction from the tyranny of the printed page into a format that better resembles how ideas are represented within our minds—and Nelson's own hummingbird style. Nelson railed for decades that the information technology industry "was in a terrible rut," fixating on making electronic documents mimic traditional paper ones in form, style, and structure, saying, "I'm mad as hell and trying to make things right."[30]

Like Douglas Engelbart before him, Nelson reached out to Vannevar Bush with a phone call while visiting the Boston area for a conference in 1968. He found his number in the suburbs by calling information, where the 31-year-old man with the hummingbird mind told the 78-year-old patrician that he had begun working on a computer system that had a lot in common with the Memex. The elder Bush had responded positively, but Nelson was put off by the sound of Bush's voice, complaining that he "hated him instantly. He sounded like a sports coach. I knew I would not follow up."[31]

Nelson's vision for hypertext was similar to Bush's, but with one fundamental difference. Nelson had described a publicly accessible, shared repository of the world's knowledge that was embodied in subsequent efforts such as the World Wide Web. Vannevar Bush had envisioned the Memex more as a personal device to aid individual research. With the Memex, a user created personalized "trails" of interests, and those trails, like the memories they were designed after, would be strengthened by continued use, or faded by disuse, while uniquely adjusting to the interests of the individual user.[32]

By the end of the 1960s, Nelson started to build a computer system that embodied his novel ideas of hypertext, called *Xanadu*. It was named after the palace in Samuel Taylor Coleridge's 1897 poem "Kublai Khan." Coleridge claims to have awakened from a "narcotic reverie" with hundreds of lines of poetry in his head after reading a book that described Xanadu, the summer palace of Kublai Khan, the emperor of China. Nelson's version of Xanadu has been called the computing world's longest-running unfinished project, with over 30 years of ultimately unsuccessful effort.[33]

Nonetheless, Nelson's unrelenting evangelizing for hypertext has inspired a number of successful projects that break the linearity of represented ideas, including Lotus Notes, Apple's HyperCard, which we will explore next, and of course the Internet. Unfortunately, instead of being proud of his influence on one of the world's most powerful innovations, Nelson laments that the Web has "one-tenth of the potential of what I could do. It is a parody. I like and respect Tim Berners-Lee,* [but] he fulfilled his objective. He didn't fulfill mine."[34] Nelson does, however, include the following item on his curriculum vitae, under the heading "Derivative work": "Arguably, the World Wide Web is derivative of my work, as is the entire computer hypertext field, which I believe I founded."[35]

HyperCard

There may be no more enabling piece of software supporting instructional media than Apple's HyperCard. It was the incarnation of Vannevar Bush's Memex and a more practical, albeit less ambitious, embodiment of Ted Nelson's Xanadu, with a bit of Douglas Engelbart's innovative sense of inventive engineering thrown in for good measure. What is most important about HyperCard is that people used it. HyperCard gave nontechnical people the ability to create original solutions that led the world to understand the value of the nonlinear, linked world that would come to represent the World Wide Web a decade later.

HyperCard's designer, Bill Atkinson, was a wunderkind and one of the original team at Apple that developed the Macintosh. Unlike Nelson and Engelbart, he didn't seem to derive his original inspirations from Bush, but when HyperCard debuted in 1987, the linked media statesman's ideas, echoed by his acolytes over four decades, were now part of the winds fueling innovation in Silicon Valley.

In the late 1970s, Bill Atkinson (fig. 3.5) was a graduate student at the University of Washington, close to finishing his PhD in neuroscience. As the personal computer revolution was coming to a head, he succumbed to its beckoning; he "got a quick E.E." (electrical engineering degree) and started a small technology business. Out of the blue, he received a call from an old friend and former professor from his undergraduate days at the University of California, San Diego. Jeff Raskin, who had been working at Apple for six months, tried to persuade Atkinson to join him at the 30-person startup. Atkinson initially demurred, but in a stroke of brilliance, Apple sent him a nonrefundable plane ticket to come visit.[36]

Atkinson was soon swayed by what he saw in Cupertino, and by Apple's ever-persuasive president Steve Jobs's three-hour pitch: "We are inventing the future. Think about surfing on the front edge of a wave. It's really

* English computer scientist Sir Tim Berners-Lee invented the World Wide Web in 1989.

Figure 3.5. Bill Atkinson (*right*) and Steve Jobs (*left*), 1984. Courtesy of Norman Seeff

exhilarating. Now think about dog-paddling at the tail end of that wave. It wouldn't be anywhere near as much fun. Come down here and make a dent in the universe."[37] Atkinson quickly became one of the key designers at Apple, developing many of the user interface elements we take for granted today (double-clicking, menu bars, selection lasso, etc.), the popular MacPaint graphics drawing program, and of course HyperCard.

Steve Jobs seemed to have a knack for saying just the right thing to persuade a prospective employee to join him at Apple on a larger mission. While wooing Pepsi's CEO John Sculley to join as Apple's president in 1983, Jobs famously taunted the 44-year-old Sculley by saying, "Do you want to sell sugared water for the rest of your life? Or do you want to come with me and change the world?"[38]

Atkinson developed HyperCard (initially called "WildCard") as a side project at Apple after the release of the Macintosh in 1984. He had John Sculley's support for the project from the onset, and insisted that Apple bundle it freely with every Mac it shipped. This allowed Atkinson to develop his vision without having to follow typical company development-cycle rules, and he could explore directions that did not have direct revenue im-

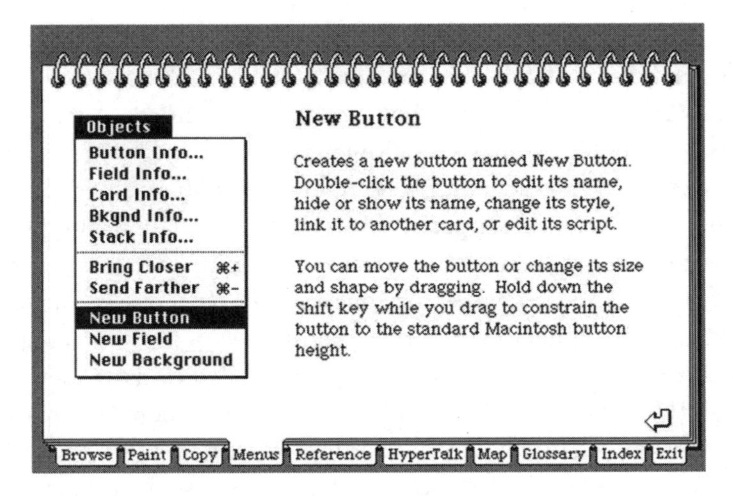

Figure 3.6. HyperCard screen. Courtesy of Ted Kaeler

plications for Apple. WildCard was renamed HyperCard and formally introduced in 1987 at the Boston MacWorld conference, to an enthusiastic audience that clamored to do the kind of things HyperCard enabled.[39]

Atkinson described HyperCard (fig. 3.6) during a 1987 television interview: "Simply put, HyperCard is a software erector set that lets nonprogrammers put together interactive information . . . We use cards that contain graphics and text and buttons . . . the cards are grouped together in stacks and you can organize it so that any card can jump to any other card. . . . The freedom to organize the information according to how things are associated with one another. . . . Cards can contain both information and interaction. That's sorta what's new here."[40]

HyperCard was met with an "undertone of disdain" from the university-centered hypertext community dominating the field at that time.[41] HyperCard required no programming experience and it became widely used even though it was not a professional authoring tool. It may have threatened some researchers' more ambitious and idealistic goals, as well as their sense of letting the hoi polloi into their inner sanctum. Apple countered, "Creating the right tool is an issue of iteration. HyperCard may not be the *right* end tool, but is *an* end tool."[42] Apple itself had some business misgivings about HyperCard, worrying that it would compete with Apple's extremely loyal developer community, but the tool struck a chord with the emerging field of multimedia.

For learning environments, HyperCard could put back what formal education had torn away—the relationships between things—and it added a sense of storytelling and narrative flow. It made improving instructional materials easier by allowing for multiple perspectives on a subject and contextualized it by linking multiple representations of the same phenomenon together. This made it possible to encode the relationships between these different representational systems that encourage learners to develop intuitions and make abstract concepts more concrete and accessible. HyperCard also challenged the author–learner relationship, making it less autocratic. Learners had much more agency to control how they moved through the material than they had with traditional, more linear media.[43]

As supportive as John Scully had been, Steve Jobs was less enthusiastic about the HyperCard project when he returned to Apple in 1997. Fabrice Florin, one of the first users of HyperCard at the Apple Multimedia Lab, said, "Steve Jobs didn't like HyperCard. I met him at a party at Bill Atkinson's house, and he wasn't big on it. When he returned to Apple, he killed it immediately. He didn't think people should be encouraged to tinker like that. . . . HyperCard was democratizing Jobs' design philosophy."[44]

But HyperCard was important for that very reason. Using HyperCard, you didn't need to be a programmer to create compelling and powerful entertainment and instructional content. It was truly enabling. That potential became even more pronounced when Apple programmer Ted Kaehler added an interface to HyperCard so it could control videodiscs. Apple's HyperCard interface designer Kristee Rosendahl later opined, "Once HyperCard was hooked into a laserdisc player, it took on a new level of importance. I don't think anybody, including Bill, really knew that it would be as dynamic as it turned out to be. It sparked and enabled a whole generation of multimedia presentations and tools that I'm not sure were fully thought out from the very beginning."[45]

HyperCard also prepared a whole generation of developers and computer users to become familiar with the ideas of hypertext and linking that were the foundation of the World Wide Web. Atkinson had some misgivings about not realizing how close HyperCard was to what the Web became. In 2002, he mused, "I have realized over time that I missed the mark with HyperCard. I grew up in a box-centric culture at Apple. If I'd grown up in a network-centric culture, like Sun, HyperCard might have been the first Web browser. I thought everyone connected was a pipe dream. Boy, was I wrong. I missed that one."[46]

Multimedia

Integrating Media and Computing

Multimedia is the term describing the process of presenting information through multiple forms of media—such as words, images, video, audio, and animation—in a unified manner on a single delivery system. While there is no proof yet to support the enticing idea that people possess individual preferred ways of learning, sometimes called *learning styles* (i.e., auditory, visual, tactile, etc.),[47] the evidence is clear that some ways of representing information are more effective than others. This effectiveness depends on using media forms that are compatible with what is being taught and with each other.

Multimedia becomes potentially more powerful when the personal computer enters the scene. Like many computer industry–driven ventures, the term multimedia has unfortunately taken on more of a marketing-driven definition that belittles its transformative potential in education. Since the days of the early PLATO and TICCIT instructional systems in the 1960s and 1970s, computers have been used to deliver multiple forms of media. But it was the rising power and affordable price of the personal computer during the 1980s that made possible the wide range of media we consume on our computers, tablets, and phones today. Apple pushed the boundaries of what was possible, and, when coupled with HyperCard, it provided some excellent models for effective multimedia applications in education.

Multimedia Theory

Drawing on the work of the philosopher Ludwig Wittgenstein, some researchers believe that multiple representations of ideas engender a deep understanding of a complex landscape that cannot be achieved by a single transversal, and that "*landscape must be crisscrossed in many directions* to master its complexity . . . and should be *revisited* from different directions, thought about from different perspectives, and so on. There is a limit to how much understanding can be achieved in a single treatment, in a single context, for a single purpose. By *repeating* the presentation of the same complex case or concept information in *new contexts*, additional aspects of the multifacetedness of the landscape sites are brought out."[48]

Decades of research suggest that, when done properly, using multiple (but appropriate) forms of media together can be more effective than any single media form alone. This evidence is borne out in the humble book, where text and images are routinely shown together; it is also achieved in videos, where moving images, still images, and sound appear jointly.[49] The psychologist Richard Mayer outlined a number of principles describing some of the characteristics of multiple-channel learning:

1. *Multimedia principle*: People learn better from words and pictures than from words alone.
2. *Split attention principle*: People learn better when words and pictures are physically and temporally integrated.
3. *Modality principle*: People learn better from graphics and narration than from graphics and text.
4. *Redundancy principle*: People learn better when the same information is represented in multiple formats.
5. *Segmenting, pretraining, and modality principles*: People learn better when information is broken into segments, when they know the names of the main concepts, and when words are spoken rather than written.
6. *Coherence, signaling, and spatial/temporal contiguity principles*: People learn better when extraneous material is excluded, when pictures and text are close to each other, and when graphics and narration are present.[50]

The Mother of Multimedia

Despite the potentially chauvinistic overtones the heading above implies, Kristina Hooper Woolsey (fig. 3.7) has proudly worn this moniker in her work evangelizing multimedia over the past 40 years. She is a California native, and her path from academia to the computer industry is a story of being in the right places at the right times, and proactively making the most of being there. By training, Woolsey is a cognitive psychologist, the first doctoral student of the legendary user-centered design guru and author of the seminal book *The Design of Everyday Things*, Don Norman.[51] His influence is easily seen in her work over the years, with an emphasis on putting the learner in control of their experience and instilling a finely

Figure 3.7. Kristina Hooper Woolsey (*left*) with Douglas Engelbart (*right*), 2009. Courtesy of Alan Levine

honed ethic that values aesthetic and well-designed products that have a sense of fun.[52]

Woolsey has a long history in appreciating the power of the geographic metaphor to simulate spatial navigation by using technology, first by working on the Modelscope projects at the Berkeley Environmental Simulation Laboratory. She then became a visiting professor at MIT's Media Lab (née Architectural Machine Group), where she worked on the Aspen Movie Project in the late 1970s. These experiences would prove invaluable to her work cofounding and directing Apple's Multimedia Lab. She once said, "Everything we have ever done in the lab, I understood in 1977."[53]

At the request of a former colleague who had just joined Apple, Woolsey gave a talk at their offices in Cupertino. Apple representatives asked if she wanted to work with them, beckoning in the familiar siren call: "We're inventing an educational computer. We want you to help."[54] Apple was the epicenter of the personal computer revolution in 1984 with the introduction of the Macintosh. Woolsey joined the educational research group, staying 14 years at various positions in the company, and finishing up as an Apple Distinguished Scientist in 1998.

Apple's Multimedia Lab grew from a need to introduce some new ideas into the engineering group that could take advantage of the upcoming increases in screen resolution, graphical capabilities, and newer mass storage options such as the CD-ROM. In 1985, Woolsey chose to look outside for some high-profile media partners to help infuse Apple's engineering-driven culture with some fresh ideas: the National Geographic Society and George Lucas's production company Lucasfilm, located just across the Golden Gate Bridge from San Francisco.

Whether it was proximity, personal relationships, the chance to work with Apple, or some combination of the three, Woolsey developed a close working relationship with the producers of the Indiana Jones movies. Unlike the *Jasper Woodbury* producers, Woolsey was able to capture George Lucas's personal attention, and not just the wrath of his legal team. Taking a cue from Nicholas Negroponte's "demo or die" attitude at the MIT Media Lab, they worked closely with the Lucasfilm on design and National Geographic for content to produce compelling demonstration projects, including *GTV: Geography Television*.[55]

GTV was an early National Geographic Society videodisc project, designed in response to an internal mandate to improve geography education for K–12 students. The program was aligned to the California State curriculum standards for history and geography, and could be used in both whole-class and small-group instruction. The goal was to move geography away from the mindless memorization of place names and their products. The interactive nature of the computer-driven videodisc made the subject much more interesting than simply memorizing state capitals and the like, and promised to "restore geography to its rightful place as an exciting, integrative way of organizing knowledge."[56]

In 1987, Woolsey and her colleague Sueann Ambron started a skunk works for Apple in downtown San Francisco, some 43 miles from Apple's Cupertino headquarters, to develop prototypes of multimedia projects using HyperCard's predecessor, WildCard. Over the course of the five years it was in operation, the lab produced a number of innovative projects that pushed the envelope for effective use of multimedia in education.

These projects included a collaboration with the Natural Audubon Society, Lucasfilm, and WGBH (a Boston PBS station) that extended the spatialization metaphor to create an interactive cabin of a naturalist detective in his quest for clues in the case of *The Mystery of the Disappearing Ducks*. The disc contained interviews with ecologists, farmers, and hunters as fodder for a mystery to be solved by the protagonist, Paul Parkranger, in a manner similar to Vanderbilt's *Jasper Woodbury* series.[57] The *Grapevine: Voices of the Thirties* project took an interactive look at John Steinbeck's classic 1930s novel *The Grapes of Wrath* and was developed in collaboration with two high school teachers.[58] The *Visual Almanac* was perhaps the lab's best-known videodisc project, with over 7,000 images and sounds that students could browse and use to build their own presentations and reports.[59]

Apple closed the Multimedia Lab in August of 1992, but its impact on promoting multimedia-based instruction, particularly with the use of HyperCard, had a far-reaching influence. In a sense, its closing was a marker of that success. The lab had defined multimedia to a generation of educators and computer users, and those capabilities were becoming mainstreamed into the overall computer experience. Multimedia capability had become a standard part of a modern personal computer, and working with images, video, sound, and animation became standard expectations from people who used them.

Digital Video

Videodisc was a game-changing technology for education. Its large capacity and ability to show 54,000 individual images, 30 minutes of high-quality video, or some combination of the two brought new capabilities to instructional designers. Videodisc's capability to quickly access any point on the disc within seconds uncoupled video from its linearly constrained roots, especially when connected with a computer and navigational software such as HyperCard.

The early television pioneers rebelled: they knew that the mechanical methods used in their time to capture and display video images could not provide the qualities needed for effective television. Instead, they employed electronic circuits to force electrons to scan vacuum-filled tubes. In our era, state-of-the-art technology is the digital computer; its ability to process information offers an effective way to use a single device to rapidly transform information in useful ways. To participate in this new digital world, video needed to make the transition from an electronic signal to a source of information. This process is called *digitization*, and it has been transforming products from almost every area of technology.

Nicholas Negroponte has described the process of digitization as converting atoms into bits, likening the consequences to a Trojan horse, with new opportunities emerging to threaten older business models (such as Napster and Uber challenging the record industry and traditional taxis, respectively).* Because "bits are bits," he said, they "comingle effortlessly. They start to get mixed up and can be used and reused together or separately. The mixing of audio, video, and data is called multimedia; it sounds complicated, but is nothing more than intermingled bits."[60]

Digitizing Video

Just like physical film and analog video, a digital video stream is made up of individual frames, each one representing a time slice of the scene. Films display 24 frames per second, and American video presents 30 frames in that same time span, also known as the frame rate The higher the number of frames per second, the smoother the video will appear. Digital video clips use frame rates from 12 to 30 frames per second, with 24 frames per second commonly used. The audio (fig. 3.8) is stored as a separate stream, but kept in close synchronization with the video elements.

Like analog television, digital video uses a "divide and conquer" strategy. But in addition to dividing the image into a series of horizontal lines, each of those lines is further divided into a series of dots, called *pixels*, and each dot's intensity and color are represented by a number. If we were to look at a frame of digital video and zoom into it, each of these discrete pixels would be easy to identify (fig. 3.9). We can visually identify each pixel according to its overall intensity and color, but that color can be represented by a number that uniquely identifies its overall value, making it easier for the computer to manipulate and store.

We can thank the phone company's work on their Picturephone as the catalyst for developing bitmap graphics (think of the scene on the transport to the moon in Stanley Kubrick's film *2001: A Space Odyssey*, fig. 3.10).

* Napster was the online file-sharing service that began the decline of CD sales. Uber threatens traditional taxicab services with cars and drivers that appear on demand from a smartphone application.

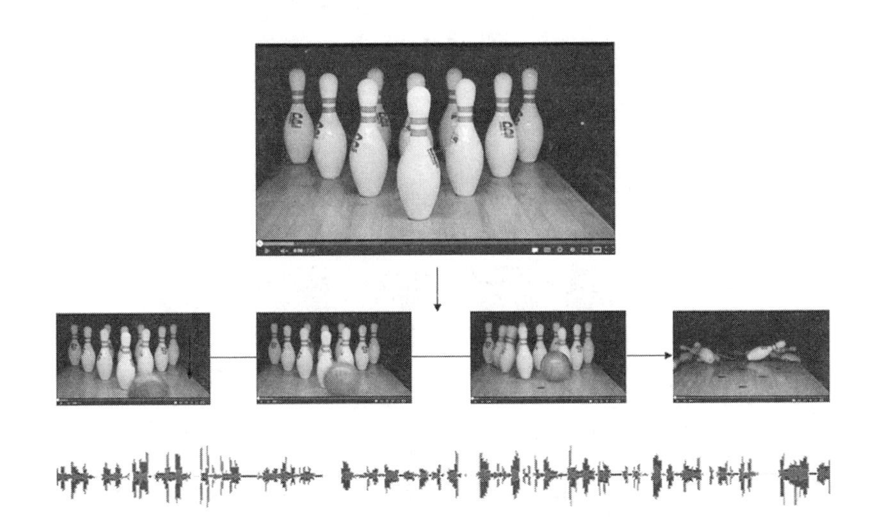

Figure 3.8. Frames in a digital video clip.

The Picturephone debuted at the New York World's Fair in 1964 with the promise to add video to everyday phone calls, but it never caught on. The first versions were essentially low-resolution traditional television systems, but ATT's research arm, Bell Labs, having invented the transistor some years earlier and being on the forefront of developing graphical computers, wanted to digitize the image process.

Bell Labs researcher Michael Noll demonstrated a system that digitally connected each pixel on the screen with a computer's memory that stored the numerical value of the pixel's visual appearance. Once digitally yoked together, any changes to the computer's video memory would be reflected instantly on the screen. Known as a bitmapped display, it is the basis of how all computers and digital television screens work today. This was a profound change in the relationship between computers and images, because images could be generated and manipulated as fast as the computer could change the numbers that represented them in memory.[61]

The amount of memory dedicated to the video display is what controls the perceived quality of the video. Representing the image in fewer dots, and therefore less memory, creates a grainier, more pixelated look. This is akin to looking at a pointillist painting whose many finely placed brush strokes are visible close up, but when viewed from a distance the

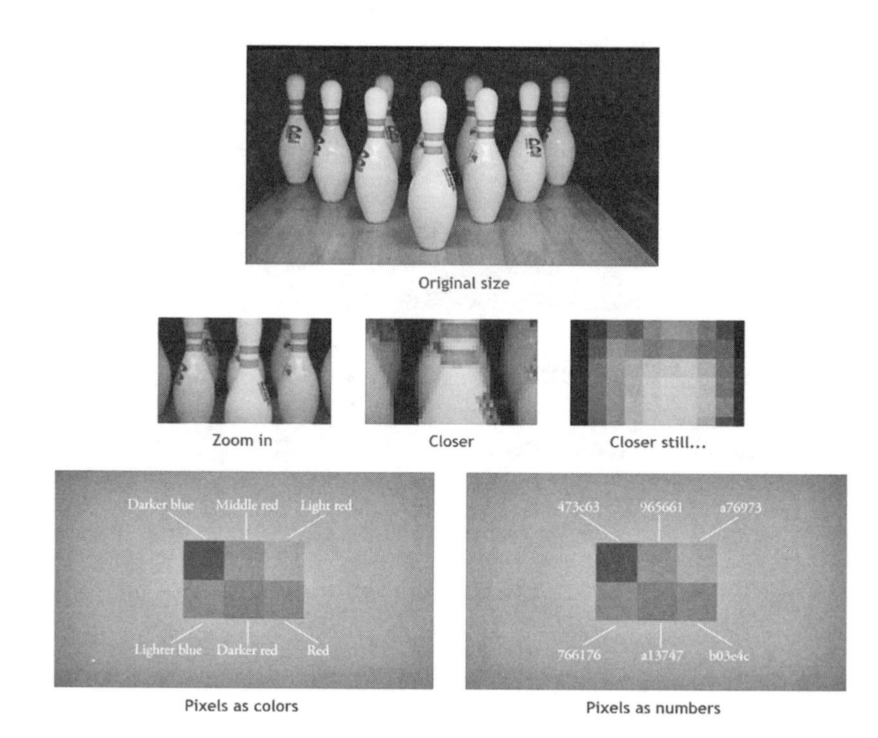

Figure 3.9. Frames into pixels.

Figure 3.10. Picturephone from *2001: A Space Odyssey*.

entire image looks smooth. Early digital videos were small, typically 320 pixels horizontally by 240 vertically (fig. 3.11). As memory became cheaper and computers faster, larger images were easily displayed with thousands of pixels across, affording the lifelike image quality we see on modern high-definition (HD) displays.

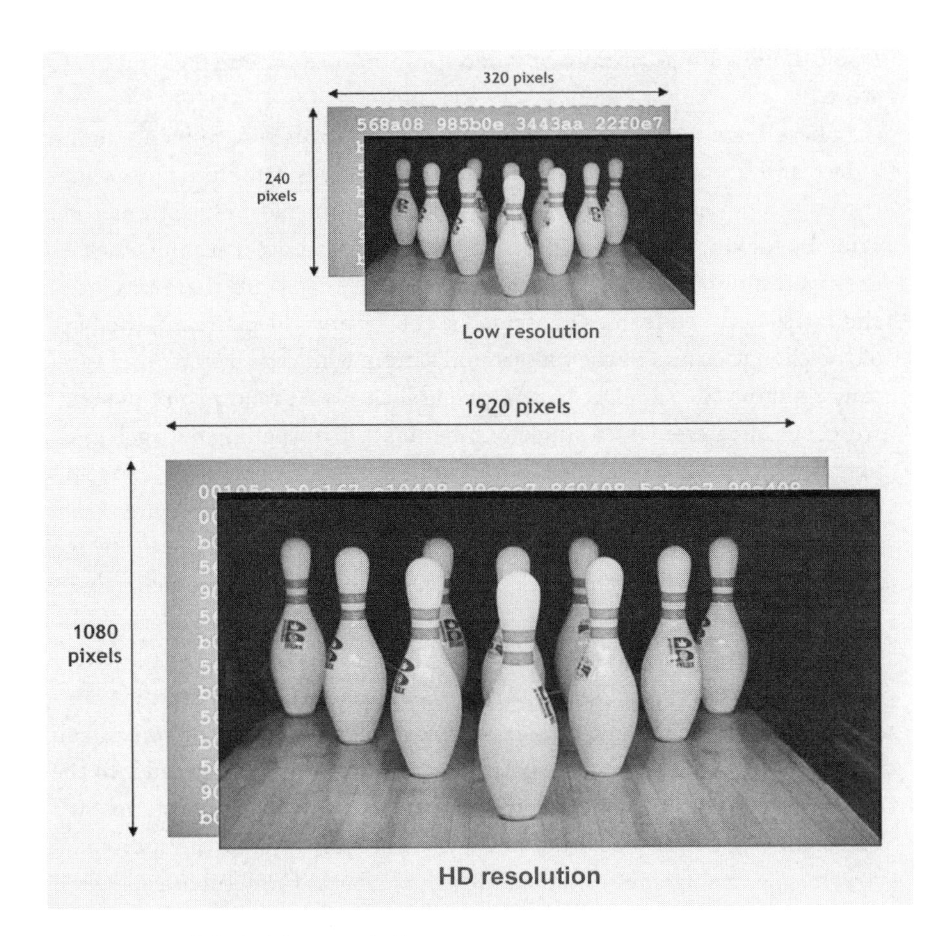

Figure 3.11. Digital video resolution.

Unfortunately, the more pixels in an image, the more space it will take to store, and those numbers add up quickly. The low-resolution video at the top of figure 3.11 contains 240 lines of 320 pixels, requiring 76,800 pixels in all. Each pixel is made up of red, green, and blue values, each requiring 3 bytes, and totaling over 200 kilobytes to store.* But there isn't just one frame in a video clip; there are 24 of them per second. Each second needs 5 megabytes to store, a full minute requiring 332 megabytes. The scale of these numbers becomes staggering at HD resolutions. With HD

* Where $240^{\text{lines}} \times 320^{\text{pixels / line}} \times 3^{\text{bytes / pixel}} = 203{,}040^{\text{bytes / frame}}$.

resolutions, a single minute of video contains a whopping 8 gigabytes to store.*

These huge numbers presented a practical roadblock to using digital video, and a number of mathematical techniques attempted to reduce the large amount of storage needed. In the end, the most efficient method came by looking at small chunks of the image and determining whether they were similar to other image chunks. As it turns out, there is a lot of similarity between frames in a video scene. In any video stream, the bulk of the change occurs in the foreground action, while the background typically remains the same. A 16×16-pixel block would require 768 bytes to store, but by referencing a single number instead of spelling out each pixel in the block, the size could be reduced dramatically. The MPEG video standard does this (among a number of other tricks) to reduce HD video from 8 gigabytes per minute to a still large, but more manageable, 100 to 150 megabytes per minute. The audio portion of the clip is compressed using a variant of the MP3 compression used on popular online music sites.[62]

Compressing individual frames is only part of the solution to deliver digital media via the computer. An overall framework is required to marshal the flow of the media data from the storage device to the screen and speakers. This framework is a software application, often bundled into the computer's operating system, such as Apple's QuickTime and Microsoft's Video for Windows, which defines a mechanism to wrap the individual streams into a single file and mediates its playback. Unfortunately, even if the underlying streams are compressed by using an industry-standard format such as MPEG, the streams are often incompatible with one another, requiring the installation of special software.

CD-ROM

The CD-ROM disc was introduced in the mid-1980s, offering an inexpensive way to distribute a relatively large amount of digital content (up to 700 megabytes) to a mass audience. Previously, delivering an application with this amount of storage required hundreds of bulky and expensive floppy discs that needed to be painstakingly loaded one by one to the computer's hard drive. CD-ROMs freed developers to use attractive

*Where $1024^{\text{lines}} \times 1920^{\text{pixels/line}} \times 3^{\text{bytes/pixel}} \times 24^{\text{frames/s}} \times 60^{\text{s/min}} = 8{,}493{,}465{,}600^{\text{bytes/min}}$.

graphics, animation, video, and audio to create compelling multimedia experiences that they could not practically deliver before.

A lot of industry hype surrounded CD-ROM at its launch, with inexpensive disc players quickly becoming standard equipment on personal computers. Microsoft's Bill Gates offered a McLuhan-esque outlook on its progress, saying, "In the same way the first TV shows consisted mainly of someone standing in front of a microphone and reading a radio script, the initial efforts in CD-ROM will not stretch the medium to its limit." He urged designers to use CD-ROM to "go beyond our traditional ways of being entertained, of learning, and of gathering information."[63]

But the CD-ROM was not the heir apparent to the videodisc. While videodiscs could be controlled by a computer, they were not themselves digital. Videodiscs stored high-quality video and audio in an analog format that was compatible with commonplace television displays. CD-ROMs, on the other hand, were true digital natives and could not play video and audio unless digitized into bits and compressed to meet the discs' limited bandwidth. The technology for effective compression still needed to be developed, and CD-ROMs could not be used for the same purposes as analog videodiscs.

The DVD (or digital video / versatile disc) took the videodisc's role when it was introduced a decade later in 1997, able to play two hours of high-quality, compressed digital video and audio from its convenient five-inch format.[64] So the CD-ROM had a relatively short life, obviated by DVDs and later by video delivery over the Internet, but it had an important role in interactive learning and in the emerging educational games industry.

CD-ROM Games

Advances in increasing storage space gave rise to a new genre of computer applications called "edutainment," which blurred the line between games and tools for learning and were graphically rich and fun for kids. Many of these programs—such as *Reader Rabbit*, *Math Blaster*, and the *JumpStart* series—employed *gamified** versions of drill and practice activities

* "Gamification" refers to the process of adding gamelike attributes to nongame activities, such as earning points and virtual "badges" for progress through the activity. It has a somewhat pejorative connotation with game developers, somewhat akin to "putting lipstick on a pig."

to learn math, spelling, and vocabulary. But a number of excellent simulation-style games also emerged, including *SimCity*, *Civilization*, and *The Oregon Trail*, which offered players a more exploratory and self-guided environment, more aligned with Seymour Papert's constructivist/constructionist approach. Parents bought these games in the hopes of both engaging their kids and giving them an edge, and the games were routinely used in computers that were beginning to populate American classrooms.

There has been a paucity of hard research on the effectiveness of edutainment programs, so it's not clear whether these applications have any real positive effects on learning. Researchers informally looked at how students engaged with *The Oregon Trail*, which provided an attractive historical simulation of homesteaders traveling by wagon train from the Missouri River to Oregon in 1848. When properly supervised, students could gain valuable insight into the pioneer life; left to their own devices, however, they tended to concentrate more on getting through the game as quickly as possible, as opposed to thoughtfully exploring the microworld of the pioneer West and its rich historical content. Students resorted to trial and error rather than working through the problems in a thoughtful way. In short, they tended to interact with the program in ways that bypassed some or all of the educational objectives; students were more focused on the game-specific (and fun) elements of the program.[65]

The sales of CD-ROM-based edutainment games had a meteoric rise, with over $500 million in sales at its peak in 2000, and an equally swift fall, with sales of only $152 million by 2004.[66] A number of reasons can account for the decline, but clearly the emergence of the Internet played a large role. Instead of having to provide all the content on the limited space of the CD-ROM, as large as it was, Web sites could deliver an unlimited variety on demand, often for free. CD-ROMs needed to be purchased in a physical store, making them less accessible and competitive with the emerging online games.[67]

Bob Stein's Voyage

As much as Ted Nelson hates the way that new technology constantly re-creates the book metaphor, Bob Stein (fig. 3.12) embraces the potential of technology to extend the book. "The book was always fundamental to me. One of the things I really liked was that the original logo for Criterion, which we designed in 1984, was a book turning into a disc. It was

Figure 3.12. Bob Stein at a conference, 2012. Courtesy of Greg Peverill-Conti

central. When I was writing the paper for Britannica, I felt like I had to relate the idea of interactive media to books, and I was really wrestling with the question What is a book? What's essential about a book? What happens when you move that essence into some other medium?"[68]

The iconoclastic Stein has been using technology to push the boundaries of multiple forms of delivery, beginning with analog videodiscs, CD-ROMs, and most recently an exploration into the future of the book itself. A confirmed Maoist, he was a founding member of the 1960s radical activist movement Students for a Democratic Society (SDS) while doing his undergraduate degree in psychology at Columbia University. He later earned a master's degree in education from Harvard.[69]

Stein became interested in using videodiscs as a way of storing the entire *Encyclopaedia Britannica* on a single disc and, like Dustin Heuston some 30 years earlier, he was able to convince its venerable publisher to fund him on a fact-seeking mission. "You have to understand, I knew nothing. I got paid, basically, to spend a year going around the country with their imprimatur, going to every lab I could, saying, 'Show me what you've got.' I wrote a 120-page paper on the *Encyclopaedia Britannica*, which suggests that the encyclopedia of the future would likely be a joint venture of Xerox, Lucasfilm, and Britannica."[70]

As it happened, Microsoft heralded the future of the digital encyclopedia by approaching the esteemed but financially struggling Encyclopaedia Britannica to license content. But the company demurred, saying they had "no plans to be on a home computer. And since the market is so small—only 4 or 5 percent of households have computers—we would not want to hurt our traditional way of selling."[71] This decision was a classic example of the *innovator's dilemma*, proposed by Harvard Business School professor Clayton Christensen to explain why market incumbants fail to adopt new innovations. They don't want new innovations to interfere with their current business activities even though ignoring them may prove disastrous.

In 1989, Microsoft purchased Britannica's nearly bankrupt competitor, *Funk and Wagnall's New Encyclopedia*, added a significant amount of compelling multimedia content, and in 1993 released it under the name *Encarta*. The new digital format was more engaging and searchable than the print version, and it forced Britannica to release its own CD-ROM version in 1995; in 2010, publication of the print encyclopedia ceased.[72]

Freshly schooled on the capabilities of videodisc, and seeing that it was not at all suitable for a text-driven application, Stein looked for other uses for the medium that better fit its strengths. While at a meeting with the president of RKO Home Video, Stein recalled, he asked about movie rights. "'So, what's the chance you would sell me the rights to *Citizen Kane* and *King Kong* for laserdisc?' He said, 'Well, they're not worth anything to us. Of course I'll sell them to you.' So I bought the rights to two of the most famous movies ever made."[73] These two classics launched his new company, the Criterion Collection, which eventually sold hundreds of videodiscs for discerning audiences.

When CD-ROMs began to appear, Stein and his wife, Aleen, founded the Voyager Company to explore how this new digital medium—which effortlessly merged text, images, video, audio, and animation—could extend older media forms. They used the challenging slogan "Bring your brain" as their mantra. Voyager grew to be one of the premier publishers for the emerging medium, with the company publishing 17 of the top 50 CD-ROM titles listed in *MacUser* magazine in 1993.[74]

Voyager produced a number of well-received titles, including a version of the Beatles film *A Hard Day's Night*, which included interviews with the

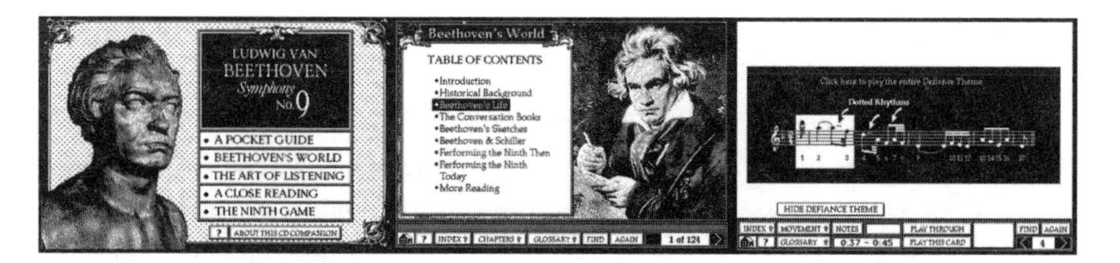

Figure 3.13. Screen shots from the Voyager Beethoven CD-ROM.

director Richard Lester, the Fab Four themselves, and live transcripts; an interactive version of Art Spiegelman's 1991 Pulitzer Prize–winning graphic novel, *Maus*, which told of his father's survival of Auschwitz; a groundbreaking digital documentary about the 1911 *Triangle Shirtwaist Factory Fire*; and, true to Stein's radical roots, a historical exploration of the period between Reconstruction and World War II that featured then-controversial articles about gay cowhands and Grover Cleveland's illegitimate child.[75]

Voyager's first CD-ROM was a HyperCard-based exploration of Beethoven's Ninth Symphony, done in collaboration with Robert Winter, a music scholar at the University of California, Los Angeles, in 1989. It contained a high-quality recording of the symphony, played by the Vienna Philharmonic orchestra, coupled with a commentary by Winter, tying in basic musical theory, conventions, and concepts surrounding Beethoven's historic and cultural context (fig. 3.13).

The Beethoven CD-ROM allowed the musical equivalent of a "close reading" of a text, where specific passages are selected, themes are identified, and attention is drawn to the voicing of particular instruments. Winter's commentary provides insight into Beethoven's life and development of the Ninth Symphony by closely linking his words with the music in both audible and musical notation forms. To make the material more engaging to younger viewers, there were a number of built-in quizzes to assess the learner's musical listening skills, as well as the ability to invent games based on the musical score.[76]

CD-ROM's 15 minutes of fame came to an end long ago, but Stein has continued his quest to extend the power of books with a new exploratory program. He recalls a conversation from 2004: "The MacArthur Foundation called and said: 'We loved the work you did at Voyager; how can we help you go back into publishing?' I said that I had no idea what it means to be a publisher right now, but if they gave me some money to start the Institute for the Future of the Book, I would think about it. And they gave me twice as much money as I asked for and no deliverables."[77] In spite of that nonmandate, deliverables—the physical artifacts of the contracting relationship, like a white paper or report—have been emerging for the past decade, and the Institute for the Future of the Book has been busy at looking at the transition of communication from the "printed pages to the networked screen."[78]

Making Sense of Hypermedia

Hypermedia involves a number of moving parts that, when joined together, provide a dramatically more interactive experience for the viewer than traditional media forms, such as film and video. Analog videodiscs controlled by a computer were the first to be able to access content in a nonlinear fashion. The ability to randomly access any portion of the media and not have to play from start to finish gave learners more self-control in their experience and made the content customizable to their own learning progress, instead of passively watching it from start to finish.

Vannevar Bush's 1945 *Atlantic* essay inspired a number of innovations that operationalized this idea of being able to dynamically link portions of content in a nonlinear fashion that was closer to his ideas about how we really think: associatively, rather than in a fixed path. Bush's ideas were refined and extended by Ted Nelson's work with hypertext, and they inspired the networked view of the world we now use daily on the World Wide Web. Douglas Engelbart's pioneering work on the mouse-driven windowed computer interface made it possible for Bill Atkinson's HyperCard authoring system to allow nonprogrammers to craft sophisticated nonlinear learning environments that made individualized learning on the computer significantly easier to deliver.

This hypermedia delivery began with simple control over an expensive videodisc player playing content on a separate display from the computer. The emergence of inexpensive CD-ROM drives, audio, and high-resolution color graphics as standard equipment on personal computers encouraged the move from analog to digital, where the various multimedia types "effortlessly comingled" with one another, providing a unified, single-screen learning environment.

The short half-life of technologies like videodisc and CD-ROM highlights some of the difficulties in adopting emerging innovations into slower-moving environments, such as education. A smaller market, such as education, relying on mainstream products has a double-edged sword. On the positive side, smaller communities can benefit from developments that would be too expensive for communities to develop themselves but have critical mass in the larger consumer marketplace. Unfortunately, however, this means that the needs of the larger audience are also addressed first, and educational issues are an afterthought at best. Consumer products tend to have a short lifespan and, like videodiscs and CD-ROMs, are more often effectively abandoned after a few years. As empowering and successful as HyperCard was in the educational arena, Steve Jobs ultimately chose to discontinue it because it had little impact on Apple's bottom line. Bill Atkinson's insistence that it be given with every Macintosh was likewise double-edged. Making HyperCard free encouraged its adoption, and gave rise to many innovative educational projects, but it also gave Apple little incentive to continue its development and distribution.

The Internet is the most recent mainstream innovation to which the educational community has attached itself, like a remora on a shark. The remora's relationship to the shark is mutually beneficial, albeit more so for the smaller fish. The educational community benefits greatly from the access to vast resources as well as the Internet's inherent nonlinear architecture that can expand on the potential instigated by the HyperCard and multimedia projects of the 1980s. Chapter 4 looks more closely at cloud-based instructional media and how Internet-based tools are causing a revolution from K–12 to college.

4

Cloud Media

Every cloud has its silver lining, but it is sometimes a little difficult to get it to the mint.

Don Marquis, 1927

The metaphor of the *cloud* is somewhat overused to describe how computers are connected using the Internet. In cloud computing, vast amounts of digital resources—such as text, data, images, audio, and video—are stored on servers and made available on demand to people anywhere in the world. The term "cloud" originated from its use in early engineering drawings, where the Internet was often drawn as a billowy cloud—an amorphous mass of free-floating information, with undefined but limitless connections and possibilities (fig. 4.1).

Making media available on demand has created a new world of opportunities for online instruction, mainly because it removed the necessity for students and instructors to be in the same place at the same time. Earlier distance education efforts, such as the popular correspondence schools, relied on slower means of delivery, such as the mail, to get content to the student and feedback returned to the instructor. The mass broadcast methods of radio and television could instantly deliver the content to the student, but they had no way for students to rapidly communicate back to the broadcaster.

Because the Internet is a bidirectional and one-to-one communication medium, it is an ideal vehicle for delivering instructional content. There are three primary ways that a student can participate in organized instruction these days: in person, online, or some combination of the two,

Figure 4.1. A typical Internet cloud illustration.

known as a *blended learning environment*. The Internet makes it easy for a wider range of students to participate, including those who work during traditional school hours, stay-at-home parents, people with physical disabilities, and residents of remote geographic locations.[1]

It's not just place that can be changed in these environments, but time as well. Unlike in-person classrooms or educational radio and television, students do not need to meet at a particular time to ingest the lecture with the instructor and other students. On-demand video technology frees students of the temporal restraints of needing to consume the content at an appointed time as well as place.

When the Internet started to become popular in the mid-1990s, the majority of people accessed it using a relatively slow dial-up telephone modem that encoded the digital signals into guttural audio sounds. With time, a variety of new communications technologies, collectively known as *broadband*, emerged to supply consumers with much faster access. These included DSL (short for digital subscriber line), cable, satellite links, and direct fiber-optic lines to the home, and allowed high-speed digital Internet access at relatively low costs. Even living in a rural area I have high-speed access to the Internet, connecting me to the world just as fast as the city folk. This new abundance has made using digital media easier to provide and access, and it has encouraged widespread use, especially for instructional purposes.

There are three basic issues that govern the quality of an Internet connection: access speed, capacity, and latency. *Access speed* determines how much information can be delivered per second, generally measured in bits. The old dial-up connections maxed out at 56,000 bits per second (about

1/2,000 of a second of a high-definition video stream), while modern broadband connections are much faster: they are able to deliver 10 million bits and more per second, easily capable of delivering an HD video stream in real time. Some Internet providers, especially mobile operators, limit the *capacity*, or total number of bits, that can be downloaded over some period of time, usually a month or a day. Finally, *latency* is the time it takes to connect with the server. This tends to be a problem only with satellite Internet providers because of the signals' need to travel miles through space and back, running up against the limitation of the speed of light.[2]

Media's Default Theoretical Framework: Behaviorism

If the designers of most educational media had any theoretical framework to support their instructional efforts, it was usually derived from behaviorist psychology, which reigned from 1900 until the 1970s. Behaviorism sought to bring a more scientific way of understanding people, focusing on observable behaviors, in a world that was then dominated by highly subjective and individualistic methods, such as mentalism, introspection, and psychoanalysis.

The psychologist E. L. Thorndike originated many of the key concepts that initially defined behaviorism at the turn of the twentieth century, among them the S-R (stimulus-response) connection and his *law of effect*. This law posited that if a learner receives a satisfactory result for an action, the learner's response tends to strengthen his or her connection with a situation, whereas an unsatisfactory result tends to weaken the connection.

In the late 1890s, Thorndike (fig. 4.2) attended Harvard University and studied under the famed psychologist William James. He began conducting psychological research on children, but he had trouble finding human subjects. So he switched from children to chickens, convinced that animals could provide insight into human learning.[3] Unfortunately, Harvard would not allow him to keep the chickens on campus, and he was forced to conduct his experiments in the basement of James's home, in a makeshift laboratory space where he used books to form the walls of the maze he ran his avian subjects through.[4]

In early behaviorism, the learner is treated as a unified whole (sometimes referred to as a *black box*), with the psychologist able to change the learner's environment and observe what happens as a result. There is no attempt to guess about what inner processes are at work with the learner, because it is not known what they actually are. The only characteristics of a learner a behaviorist can say with confidence are what they can actually observe.[5] It is easy to see how this perspective pervades today's testing-driven K–12 classroom environment.

Thorndike's ideas would be further refined decades later by another Harvard psychologist, B. F. Skinner (fig. 4.3), as a result of unbridled enthusiasm arising from his own successful experiments using pigeons. He proposed new ways to explain learning through processes he called *operant conditioning* and *reinforcement theory*. Skinner's experiments with pigeons showed that if a learner received satisfactory results of a given action it resulted in a stronger connection with the action, while unsatisfactory results tended to result in a weaker connection. Using this basic premise, complex concepts could be taught by breaking them down into a sequence of smaller steps that were reinforced. Because learning occurs when desired behaviors are systematically reinforced, Skinner theorized that learning could be accomplished by *successive approximation*, or *shaping*, where students are directed through the content by taking many small steps, each step requiring some response from the student.[6]

Plato, in his story about Socrates teaching a geometry problem to his friend Meno's slave, presaged this idea. In "The Meno,"[7] the philosopher Socrates successfully leads an uneducated boy through solving the Pythagorean theorem by asking him to respond to a series of small questions, leading him toward the solution. Skinner's steps were designed to be as small and as directed as Socrates's questions. Each step built on what the student already knew and expanded upon that toward the next frame. While the steps were small, a skillful instructor could create a gradual progression that chained the steps together to create more complex understandings over time.[8]

This idea of building slowly on what the learner already knows had been described in the 1930s in the work of the Russian developmental psychologist Lev Vygotsky (fig. 4.4). Vygotsky's zone of proximal development theory can be thought of as series of enclosed circles (fig. 4.5), the outermost circle containing the material to be learned, the innermost circle containing the material the student already knows, and the space between the two circles containing the content the student can learn with some

Figure 4.2. Edward L. Thorndike, circa 1912. Courtesy of *Popular Science Monthly*

Figure 4.3. B. F. Skinner, 1950.

guidance. Vygotsky believed that the most efficient place for learning occurred in this middle area, or the zone of proximal development.[9]

The emphasis on observable behavior from Skinner's experiments on laboratory animals gave him confidence that the results would lend insight into how people learned. In his view, "teaching is a matter of arranging contingencies of reinforcement under which students learn."[10] The concepts of shaping and reinforcing behavior by rapid and continuous feedback have set the theoretical foundation in instruction for over a half a century now, especially instruction involving technology and media.

Streaming Media

Most types of Internet-based files need to be downloaded completely before they are viewable. This process can take an especially long time for digital media files, which can be very big. Even a highly compressed 10-minute video clip takes up almost a gigabyte to transmit. This could take somewhere between 1 and 10 minutes to download, depending on the connection speed, and that waiting time puts up a practical barrier to

Figure 4.4. Russian psychologist Lev Vygotsky.

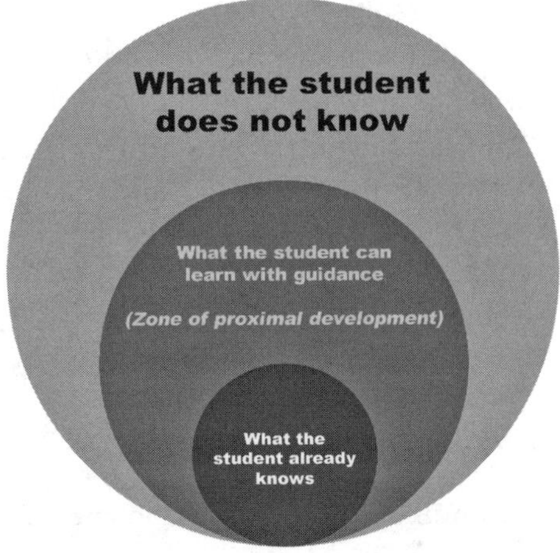

Figure 4.5. Vygotsky's zone of proximal development.

effectively using video for instruction. Luckily, there is a workaround for this barrier: in some cases, a user may only need to see a small portion of the large file at any given time, so the computer can be busy processing the file even as the user is watching it. The initial delay is thus very short, which makes watching large media files infinitely more practical.

This clever multitasking process is known as *streaming*, and it has enabled the flexible use of all kinds of dense media, from audio and video to complex 3D simulations, to be effectively used in an on-demand, just-in-time manner over the Internet. Streaming also offers some sense of security to the providers of the media, because the entire file is never saved on the person's computer. Even as the user downloads and views each new portion of the clip, the complete file is never stored, so that media cannot be as easily shared as nonstreamed files, such as MP3s. The need to stream large content versus downloading was more acute in the days of telephone dial-up for Internet access, with bandwidth being more limited, but the issue of a the large initial delay still exists in the broadband world and is effectively solved with streaming.

RealNetworks

If someone had asked Rob Glaser as a child what he wanted to do when he grew up, he might have said sports announcer for the Mets or the Yankees. Not particularly athletic and a bit overweight, Glaser concentrated on baseball statistics, memorizing many of the facts in the *Encyclopedia of Baseball* rather than actually playing the game himself. After graduating from Yale in 1983 with a bachelor's degree in computer science and a master's degree in economics, he heeded his generation's call to join the emerging personal computer revolution by heading off to Washington State and joining Microsoft. His timing was good. Microsoft was not yet the giant public company it is today, and he ultimately became one of the 10,000-plus Microsoft millionaires[11] (and part owner of the Seattle Mariners baseball team).[12]

Glaser left Microsoft in 1994 to start what would become known as RealNetworks, a company that developed software to stream audio and video content over the nascent Internet. Seizing on his lifelong passion for baseball, the company proved their technology by being the first to stream a live event over the Internet, a game between the New York Yankees and the Seattle Mariners. By the turn of the century, Glaser's company would ultimately deliver 85 percent of streaming content over the Internet. Eventually RealNetworks would fade into obscurity, as more and more companies created streaming products of their own, including Apple and Microsoft, but these pioneering efforts laid the groundwork for the rich array of streaming media that millions routinely view today.[13]

In the mid-2000s, four independent factors converged to enable a new way to store and view rich online media: First, inexpensive digital video cameras made it easier to directly capture good-quality video streams from webcams, cameras, and even mobile phones. Second, inexpensive hard-disc storage increased in capacity, so that storing these large video files was no longer a major issue. Third, a number of inexpensive and powerful video tools became available, including free editors bundled with the two major computer platforms, Apple's iMovie and Microsoft's Movie Maker. Finally, fast broadband Internet access made distributing and accessing high-quality media, on demand, practical.

YouTube

The streaming media technologies provided by RealNetworks, Apple, and Microsoft enabled people who already had Web sites to make rich media available, but the average computer user found it difficult to take advantage of these new online capabilities. It would take another Silicon Valley firm founded by three young entrepreneurs to create a platform that could make sharing and using video the ubiquitous experience it is today, and make possible the host of instructional options currently available.

Chad Hurley (fig. 4.6A) grew up in the suburbs of Philadelphia, with strong interests in technology, business, Web animation, and art. After graduating college with a degree in graphic design in 1999, he read about a new early-stage company, or *start-up*, in Silicon Valley in *Wired Magazine* that would make it easy for people to transfer money using their computers. He applied and became one of the first 20 employees of PayPal, after redesigning the company's logo as proof of his design abilities.

PayPal was a convivial work environment, and while there, Hurley met a software engineer named Steve Chen (fig. 4.6B). Born in Taipei, Chen and his family immigrated to Chicago when he was young. He studied computer science at the University of Illinois at Urbana-Champaign, but was lured to join PayPal by one of its founders just two semesters short of graduating. Another colleague to join Hurley and Chen was an East German émigré from St. Paul, Minnesota, Jawed Karim (fig. 4.6C), also an Illinois computer science student, dropping out in 2000 to follow that same PayPal executive's enticing beckon.[14]

Hurley, Chen, and Karim became fast friends—the Three Musketeers of PayPal—and when the company was acquired by online auction giant eBay for $1.5 billion in 2002, they, like Rob Glaser of RealNetworks, became instant millionaires. In their early 20s and too young to retire, they plotted their next venture. On Valentine's Day 2005, they decided to create an online dating site they named YouTube! with the slogan "Tune in, hook up." It was to be video-based version of a then-popular Web site with strong sexist overtones, HOTOrNOT.com, where one could quickly scan pictures of people and rate them hot (or not).[15]

Oddly enough, the trio had trouble attracting women willing to post videos of themselves for this kind of scrutiny, and Karim offered a $20 bounty per video on the online classified Web site Craigslist. Although there were no female takers of this specific offer, people of all genders began to upload their non-dating-related videos to the YouTube site. The founders quickly recognized the more general business opportunity and abruptly changed course.[16]

Figure 4.6 (A) Chad Hurley, 2009. Courtesy of Remy Steinegger; (B) Steve Chen, 2009. Courtesy of Joi Ito; (C) Jawed Karim, 2008. Courtesy of Robin Brown

YouTube grew in popularity over the next two years, and many people were surprised when the search giant Google acquired them for $1.65 billion in 2006. Unlike many corporate acquisitions, YouTube flourished within Google, making it easy for anyone to upload videos to the site and create collections, called *channels*, to distribute them by. The site has grown to be the dominant provider of video on the Internet, to the point that both the Vatican and the US Congress had their own YouTube channels by 2009.[17] As of 2015, YouTube had over a billion users uploading over 300 hours of video *every minute*.[18]

There are other online video-hosting sites, including Vimeo and Flickr, but YouTube remains the dominant provider in the consumer world. Because the site is supported by advertising, YouTube videos are free to upload and view. Educators, especially those in K–12, are wary of using YouTube because of the interference and potential bad influence these ads could have on kids, but even so YouTube is still the leading provider of educational content. Google and many other third parties (including the University of Virginia's Qmedia tool*) provide free, easy tools for adding titles, graphics, annotations, transcripts, and quizzes to videos.

Alternatives to the Video Camera

The modern age of computer-augmented media provides a number of alternative ways to create rich instructional media content without ever picking up a camera to record a scene. These techniques include narrated PowerPoint presentations, computer-generated simulations, architectural walk-throughs, and, most recently, screencasts. These techniques can be mixed and matched with camera-generated media, and with each other, to provide effective and compelling instructional sequences.

The most commonly used of these methods is the often-reviled PowerPoint / Keynote slide presentation with a soundtrack containing a lecture or some other vocal narration. These presentations are popular because they mimic the structure and experience of the traditional classroom lecture. Also, instructors typically already have presentations available from their lectures and are familiar with creating them. Done properly, these kinds of presentations can be very effective instructionally and easy to

* Qmedia is a freely available instructional video-authoring tool for YouTube, Vimeo, and HTML5 videos. See www.viseyes.org for more information.

produce. The tools for making them from Apple, Microsoft, and Google are easy to use, and most have internal options that can add narration and save the combined audio/video presentation as a digital file that can then be uploaded online to iTunes, YouTube, or another Web site for student viewing.

Specialized software packages can create movies of computer-generated simulations of processes such as chemical reactions, animated flow charts, data visualizations, and walk-throughs of 3D architectural and historic scenes. These walk-throughs allow people to see reconstructions of distant or past environments, such as Pompeii and ancient Rome, or renderings of buildings and environments that never existed in reality. These movies can have audio commentary added, just like PowerPoint presentations, and are easily combined with other camera and non-camera-based media sources.

Screencasting is the most recent method that instructors are using to create effective instructional video without resorting to using a video camera. Screencasts are digital recordings of a computer screen, as if a camera was pointed at the screen, but the videos are pristine copies of whatever appears on the screen in real time. A number of companies offer easy-to-use software that digitally captures everything that is displayed on the screen, and creates a compressed digital video clip complete with audio narration recorded by a microphone. The video can be uploaded to a Web site or video-sharing site like YouTube or Vimeo for viewing.[19]

The technology journalist Jon Udel collaboratively coined the term *screencasting* in 2005 by asking readers of his *InfoWorld* column to come up with a catchy-sounding moniker for this emergent media form.[20] Software companies extensively used the technique to demonstrate their software remotely, without the need to download and install software applications. Educators quickly saw the potential and used it in combination with desktop drawing applications to provide the digital equivalent of a chalkboard. TechSmith's Camtasia (fig. 4.7) is the best-known commercial application, allowing an environment for finely editing screencasts while selectively zooming into portions of the screen for emphasis and adding graphical annotations, but there are many freely available alternatives, including TechSmith's own Jing software.[21]

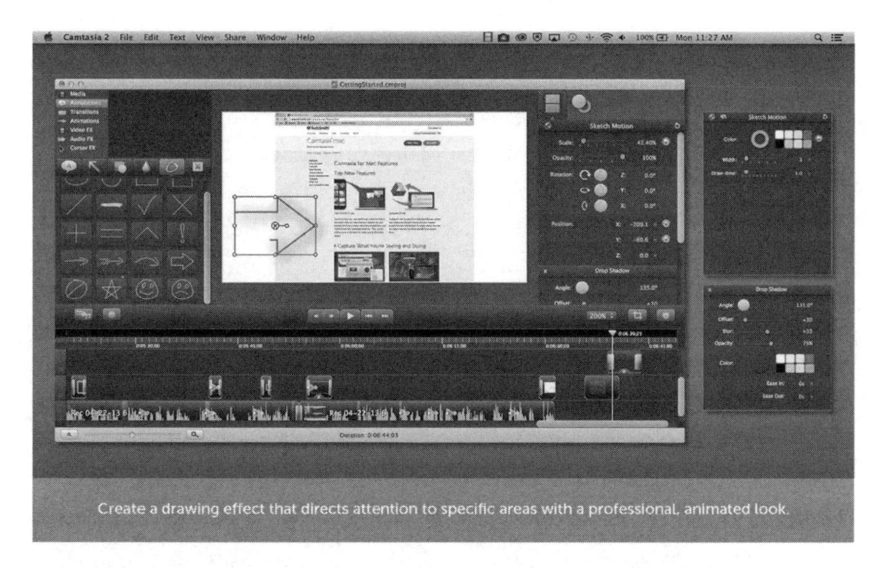

Figure 4.7. TechSmith's Camtasia tool. Courtesy of TechSmith

Recorded Lectures

In the spirit of media theorist Marshall McLuhan's observation that emerging media forms tend to replicate the style of the media that had come before them,[22] the most popular early use of video in education is recording the classroom lecture as if it were a theatrical play. Typically, a fixed camera is placed at the back of the lecture hall and the class is recorded in its entirety. The video is then made available to students as a replacement for, or in addition to, the live experience, acting as a safety net for students needing to review material or make up for missed classes.

The economic allure for recording lectures is one of economy of scale; one could produce a lesson at a fixed cost and present that experience over and over again, presumably at a lower incremental cost each time, as compared with a live instructor who must constantly deliver the lecture at the same cost per performance. Today's massive open online courses (known as MOOCs) and eLearning efforts seek the same economic and productivity goals: record the lectures of professors from elite universities once and deliver them many times to a worldwide audience of thousands of students.

Unfortunately, recorded lectures offer no real-time feedback from the students to the instructor to gauge their levels of engagement and understanding. In a physical environment, even mediocre instructors teaching impersonal 300-person introductory lecture courses are able to get a sense of student response. The current-day eLearning initiatives dismiss this reality and stake claim to other advantages, such as improved instructor quality, enhanced visuals, and the ability to replay content.

The Massachusetts Institute of Technology and other elite universities began freely posting videos of classroom lectures online—along with course syllabi, PowerPoints, and often class readings—for anyone in the world to freely access. Likewise, Apple Computer provided a clearinghouse for course material with their iTunes U site, which allowed universities to distribute lectures to students through iTunes.

OpenCourseWare

MIT took a bold step in 2001 by making academic content from almost all of its university classes freely available to anyone over the Internet. The school committed to spending up to $100 million over the span of 10 years to support the OpenCourseWare (OCW) initiative, which includes not only the videos of lectures, but also lecture notes, problem sets, simulations, PowerPoints, and even exams. This spirit of sharing runs deep in the university's DNA; MIT is one of the pioneers of the open-source software movement. One of OCW's founders, Hal Abelson, said, "In the Middle Ages people built cathedrals, where the whole town would get together and make a thing that's greater than any individual person could do and the society would kind of revel in that. We don't do that as much anymore, but in a sense this is kind of like building a cathedral."[23]

MIT's offerings have had a big impact in providing educational materials throughout the world, with over 2,200 courses from MIT and other institutions now freely available to its 2 million monthly users as of 2014. The original target audience was educators, but it turned out they constituted only 9 percent of OCW users, and they often mix and match the resources found on OCW to create their own custom online courses. The vast majority of OCW's users are students (42 percent) and self-learners (43 percent), who use the resources for personal enhancement, reviewing basic concepts, and keeping current in their fields.[24]

OpenCourseWare was an innovative approach in the face of pressures during the dot-com era's abundance of freely available Internet resources, but it would be hard to say that the state of the art in distance learning was much advanced over earlier initiatives. The free and universal admission provided by the new Internet-based offerings made access much easier than the commercial correspondence schools of yesteryear, with their reliance on the slower postal system. The ability to choose the time to view the lectures was a big advantage over the lockstep approaches of broadcast media, but the videos made available by MIT and others were typically only "filmed plays" (fig. 4.8) of the university's classroom lectures that were simply made more broadly available.

Conference Lectures

Universities aren't the only ones capturing live lectures and distributing them online. Major industry and academic conferences routinely record their speakers and offer them for viewing online. These recordings are typically more elaborate than a typical classroom lecture offering, with active camera work from multiple angles and the speaker's slides appropriately edited into the live video. Most are distributed as full audio and video streams, or just audio only, and they often provide a wealth of easily accessible information. Many are offered to the public at no cost, but some organizations delay their release so as not to compete with the live event.

Retired technology executive Douglas Kaye pioneered the idea of aggregating recorded conference lectures in 2003. He distributed them under the banner of The Conversations Network, which offered access to thousands of lectures and speeches about information technology, design, cognitive psychology, media, and social change. The Web site is no longer active, but the Internet Archive still hosts thousands of hours of fascinating hours of MP3 audio recordings.[25]

The producers of the popular Technology Entertainment and Design Conference (TED), started by the irascible information architect Richard Saul Wurman, have long made their presentations freely available online. These TED Talks are expertly delivered and well-produced speaker presentations, delivered at the organization's prestigious and costly conferences. The organizers recently launched an initiative called TED-Ed that makes it easy for educators to mix clips found on YouTube from the conferences and elsewhere, as well as local TEDx franchises.[26]

Figure 4.8. Frame from John Kuttag's Introduction to Computer Science course at MIT.

iTunes U

As the pioneer and leading manufacturer of today's audio players (first the iPod and later the iPhone), Apple Computer was quick to recognize the potential for aggregating the growing body of educational media content. In 2007, they launched a new feature on their popular iTunes music player software called iTunes U, which provides a simple interface to manage a collection of media objects for use in instruction, not unlike the learning management systems such as Blackboard or Sakai.

iTunes U rapidly became the largest provider of educational audio and video content, with over a billion downloads between 2007 and 2013. Educational institutions ranging from elementary schools to Ivy League universities have uploaded thousands of hours of lectures, demonstrations, simulations, and other instructional materials to support their existing students and to share content with others throughout the world.[27] The bulk of this content is freely available, but schools can opt to control access via passwords or to charge a fee (which is of course shared with Apple) for individual pieces of content.

The most recent version of iTunes U contains a full set of class management tools that can compete with any of the commercial learning management systems on the market. The iPad-based application makes it easy for an instructor to supplement courses with all kinds of media content, from annotatable PDF documents to video clips, as well as create and grade student assessments such as quizzes.[28]

The Flipped Classroom

Instructors have traditionally used video in two ways: to deliver instruction remotely (eLearning) and to show films and videos as a respite from lecturing. Two high school science teachers have been active instigators of a new trend in using video in the classroom: as an information resource that is watched *before* class to prepare students for richer in-class student–teacher interactions, such as problem-solving activities and discussions. This method upends the traditional classroom dynamic, where students listen attentively to instructor lectures in class and subsequently try to solve problems at home.

In 2007, Jonathan Bergmann and Aaron Sams noticed that some of their chemistry students were coming into their rural Colorado classroom with unsolved homework problems. On closer examination, it turned out that the incomplete work was not from a lack of effort, but rather that students were getting stuck on the concepts that had been theoretically illuminated during the previous day's lecture. To help them, the teachers created a series of inventive videos that combined lectures, graphics, and video clips to illustrate the phenomena being studied. They posted the videos on YouTube and asked their students to watch them at home, freeing up precious class time during which they could interact and get help from the instructors.[29]

The duo called their technique the *flipped classroom,* and that catchy label has helped spark its widespread adoption in the classroom. A 2012 survey showed that over 3 percent of the teachers surveyed were already using video to flip their classrooms, and 20 percent expressed interest in learning how to implement a flipped classroom model.[30] The availability of easy-to-use tools for video creation and distribution, such as Camtasia and YouTube, has removed many of the technical and economic barriers in recent years, not to mention improved Internet access for students.

Educators in the humanities have scoffed that the flipped classroom is not really a new pedagogical technique. English instructors have been assigning books to students to read at home and actively discuss in class for decades, if not centuries. But the STEM disciplines (science, technology, engineering, and math) have clung to the more traditional didactic lecture model and seem to be ripe for this technique, and flipping many kinds of STEM courses has been successful.[31] But preliminary research suggests that not all subjects are good candidates for flipping: introductory classes and ones that possess ill-defined problems—statistics, for example—tend not to be well received by students when flipped.[32]

Effective use of the flipped classroom model relies not just on the procedural change of the order and place of the instruction and problem-solving activities, but also on the adoption of more constructivist pedagogical teaching styles. These are typified by the student-centered theories of Jean Piaget and Lev Vygotsky that encourage the student to construct his or her own learning, with guidance from the instructor if needed. What little research that has been done to date shows little difference between whatever effect arises from flipping the classroom and traditional lecture-based instruction.[33] Even if studies in the future show some gains, it will be difficult to parse the contribution of the flipping from the more constructivist pedagogical techniques needed to properly implement it.

The Khan Academy

The Khan Academy is perhaps the best-known provider of instructional videos online and was as influential to others in encouraging the use of video-based instruction as Vannevar Bush was to the development of hypermedia. The nonprofit company was founded in 2008 by Salman Khan (fig. 4.9), who gave up his lucrative day job at a hedge fund to produce short educational videos hosted on YouTube to help tutor students in subjects such as algebra.[34] As of 2015, the Khan Academy has issued over 600 million of these lessons, ranging from basic math to physics, biology, economics, computer science, and more. Over one million teachers use the site, assigning the videos to augment classroom activities and provide remedial instruction.[35]

Khan came to instruction through the back door. While attending a family wedding in 2004, he happened to talk with his 12-year-old cousin, who was struggling with math. Nadia was a straight-A student in her New Orleans prep school, but she had done poorly on an exam that would decide her future academic pathway. It was clear to Khan that Nadia was

Figure 4.9. Khan Academy founder Salman Khan. Courtesy of Kate Mason

bright and motivated, but she needed some one-on-one help, so he offered to tutor her remotely from his home in Boston. They used the telephone and a computer program that allowed them to draw freehand on the same virtual screen, some 1,500 miles apart. Those tutoring sessions provided a laboratory for Khan to work out some of the ideas that would ultimately form his now-famous instructional video lessons.

Nadia retook her math placement exam and passed with flying colors. By 2006, Khan was tutoring a number of his family members, but he found the logistics of scheduling the synchronous phone meetings to be difficult. A friend suggested he use Google's new video-sharing service, YouTube, to host his videos alongside those of piano-playing cats. At the time, YouTube limited the length of videos to 10 minutes, but this turned out to be the ideal length for a single learning session, as later suggested by cognitive researchers. He wrote some simple math quizzing software, and the bones of the Khan Academy were formed.[36]

As it turned out, Nadia liked these videos better than the live sessions with her older cousin because she could review them any time she wanted, scrolling to the parts she still had problems with and skipping the things she already knew. Khan recalled, "She basically said, 'I like you better on the video than in person.' The worst time to learn something is when someone is standing over your shoulder."[37] Khan's insight flies in the face of the educational psychologist Benjamin Bloom's famous observation about the effectiveness on one-to-one tutoring.

Bloom found in a series of studies that one-to-one tutoring could achieve a two-sigma improvement over traditional classroom teaching. (A sigma represents the standard deviation, so a two-sigma gain is the equivalent of jumping from a score of 500 to 700 on the SAT test; see fig. 4.10). Bloom's doctoral students compared three different basic teaching styles: conventional classroom instruction, mastery learning techniques, and one-to-one instructor-student tutoring. The students were randomly assigned to one of the three styles, and Bloom found that they did best in the tutoring group. He went on to perform a meta-analysis* of a number of previous studies and confirmed his findings. Tutoring was indeed the most effective instructional method.[38] Unfortunately, one-on-one tutoring is as expensive as it is effective. For this reason, matching the effectiveness of one-to-one human tutoring has been the gold standard of educational technology. Bloom didn't compare the efficacy of video instruction in his seminal study, but other researchers have found video to be about halfway between traditional classroom instruction and tutoring in terms of its effectiveness.[39]

Math Instruction

As of 2015, the Khan Academy is the most successful provider of instructional videos on the Internet. The numbers are impressive. They claim almost 15 million registered learners, 300 million lessons, and 3 billion exercise problems delivered. They have branched out beyond mathematics, with hundreds of lessons in science—including biology, chemistry, and physics—as well as history, art history, and a big effort in computer science and programming.[40]

The most developed content area at the Khan Academy Web site remains mathematics, and they have designed a series of grade-structured curricula targeting students from kindergarten to high school. The over-

* A meta-analysis is a statistical method comparing results from a variety of different studies to identify trends and patterns from all of the results as a group.

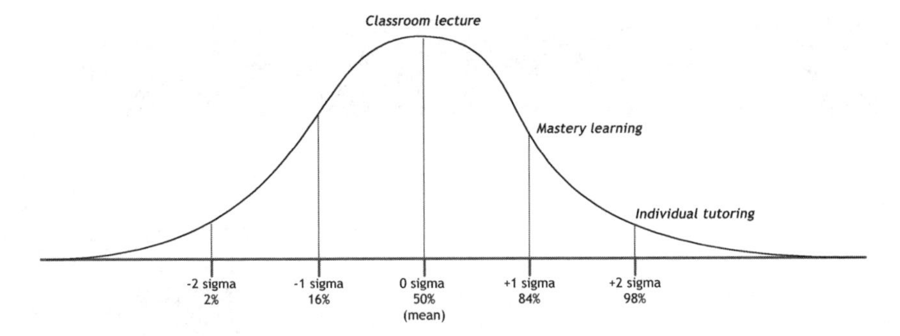

Figure 4.10. Bloom's two-sigma chart.

all curriculum down to individual lessons has been carefully aligned to the Common Core standards,* which define an evidence-based set of educational standards in math and language arts and have been adopted by a majority of states in the country.[41]

Each curriculum begins with a "mission warm-up"—essentially a pretest that assesses preexisting knowledge. Students must answer a series of questions that gauge their understandings. They can request hints if they cannot correctly answer the questions, or skip them and say they have not yet learned the material (fig. 4.11). At the end of the test, students receive a summary of their progress showing what content has been mastered and what still needs practice. The software then directs the learner to the parts of the curriculum that require more practice through a series of quizzes or practice sessions.

Practice sessions are much like the warm-ups, but they have videos (fig. 4.12) that provide guidance. These videos are typically screencasts of the instructor drawing the problem as if he were in front of a blackboard, carefully narrating his actions as he goes along. Students can watch the videos as many times as they like, at play speeds ranging from half to double time. They can respond to questions by filling in a value or dragging an onscreen widget and get immediate feedback whether it is correct or not.

The lessons promote mastery learning by requiring the learner to correctly answer five problems in a row. In most lessons, it is not possible to

* The Common Core State Standards Initiative is a controversial national content standard sponsored by the National Governors Association to standardize the content students should know at each grade.

Figure 4.11. A sample Khan "mission warm-up" test. Courtesy of the Khan Academy

move forward without supplying the correct answer. Mastery learning is particularly valuable when parts of the curriculum are heavily dependent upon the preceding lessons, as in mathematics, science, and engineering. Some mathematics educators believe that the inability to fully master the concept of manipulating simple fractions in elementary school inhibits a student's chance of success in higher-level courses, such as algebra.

The educator Henry Clinton Morrison is credited for popularizing the term "mastery learning" while he headed up John Dewey's University of Chicago Laboratory Schools in the 1920s, saying, "When a student has fully acquired a piece of learning, he has mastered it. Half-learning, or learning rather well, or being on the way to learning are none of them mastery. Mastery implies completeness; the thing is done; the student has arrived. There is no question of how well the student has mastered it; he has either mastered it or he has not."[42]

SAT Test Prep

Getting the highest score possible on the Scholastic Aptitude Test (SAT), the College Board's gatekeeper to college admission, has fueled an $861 million SAT preparation industry for achievement-oriented high

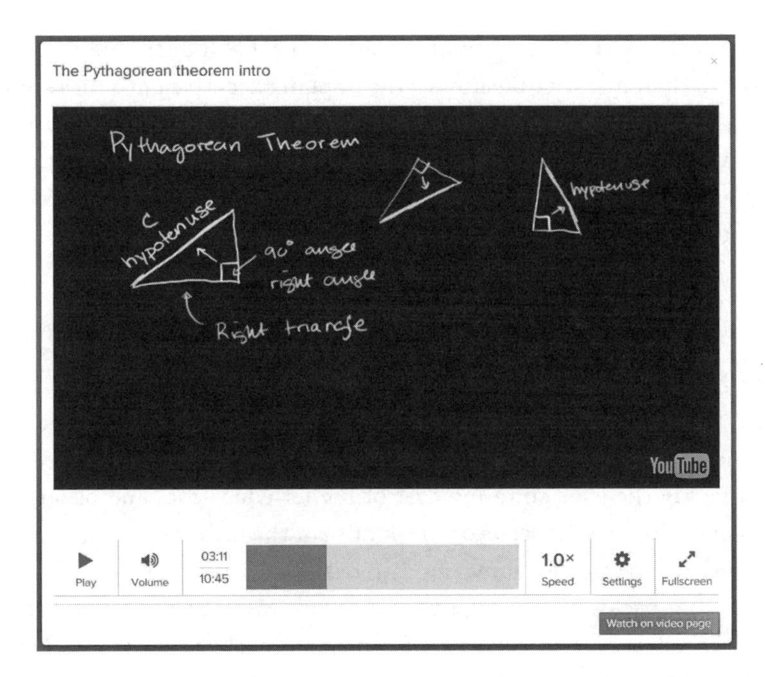

Figure 4.12. A sample Khan practice session video. Courtesy of the Khan Academy

school students and their equally anxious parents.[43] For around $1,000, companies such as Kaplan and Princeton Review promise to significantly raise students' scores, adding fuel for ambitious students wanting to get a leg up on their peers.

Recently, the Khan Academy has teamed up with the College Board to provide free online resources to level the economic playing field for test takers. They have created a collection of instruction videos that provide personalized test preparation in the same content areas as the commercial companies for the all-important SAT test. These include interactive practice tools, an emphasis on personalization and feedback, and access to four full-length official SAT practice tests. Like the Khan Academy's math curriculum, students are pretested and offered a personalized pathway through the videos, with timely feedback and many practice opportunities. The results from the official practice exams channel students to materials they need more help with.[44]

The test preparation companies should probably fear this new competitor to their lucrative business. One Kaplan executive put a nice spin on things, saying that new online resources are "a recognition that test preparation works, and works pretty darn well. Many students will want something extra to set themselves apart,"[45] preying on the innermost fears of nervous students and their parents.

MOOCs

If one pays any attention to the press, from opaque and seldom-read educational journals to the *New York Times*, the premiere educational media at the beginning of the twenty-first century is the massive open online course, or *MOOC*. There are those who view MOOCs as the savior to managing the ever-spiraling cost of higher education, and others who see them as sowing the seeds of the demise of the university as we know it. The truth, of course, lies somewhere in between.

Working backward through the acronym, MOOCs are *courses* taught *online*. The classes themselves are asynchronous, but they typically have defined start and end dates, cover a specific topic, and use the Internet to deliver curricular content to learners. MOOCs give assessments—sometimes within the video-based content, sometimes summatively—and they may offer a series of problems to be solved. MOOCs are *open* to anyone, typically at no cost, and they can involve a *massive* number of simultaneous learners, often in the tens of thousands.

The venerable *New York Times* called 2012 the "year of the MOOC," as MOOCs burst onto higher education in full force.[46] The interest was fueled by large amounts of Silicon Valley–based venture capital and publicity in both the popular and educational press, but they had a namesake some five years earlier in Canada. These Canadian MOOCs shared the same four-letter acronym and their use of the Internet, but not much else. MOOCs in Canada were facilitated by a wide range of available technologies, including e-mail, Facebook, RSS (or rich site summary), blogs, wikis, WordPress, Drupal, Twitter, and the open-source learning management system Moodle. Canadian MOOCs emphasized learner control over the interactive process, held weekly synchronous meetings with the facilitators and guest speakers, and each week e-mailed a newsletter summarizing the activity in the social media used. A high importance was placed on learner autonomy and the use of social media to self-organize and make sense of complex subject areas, in addition to the importance of learners' abilities to share their understandings with the class by creating digital artifacts, such as blogs, videos, images, and concept maps.[47]

The newer American MOOCs better resemble the mail-based correspondence schools of a century ago than their more innovative namesakes to the north. They typically define a specific set of content comprising a course of study and instruct using short video clips to deliver intimate—but one-directional—tutoring over the Internet. Learners have an opportunity to test their understandings through quizzes and other assessments, often directly embedded within the video clips themselves. Some of the more progressive providers offer interactive environments, such as simulations and test beds to both instruct and assess understandings. Homework is assigned and graded using a variety of automated methods, and an overall grade for the course is given at the end.

The modern MOOC era officially began on October 10, 2011, when Stanford computer science professor Sebastian Thrun and Google's director of research, Peter Norvig, offered the course CS221: Introduction to Computer Science to over 160,000 registered learners, resulting in what some have called "the online Woodstock of the digital era."[48] The graduate-level course was simultaneously taught in person at Stanford and using video over the Web. Over 200 students came to the first class, but that number quickly dwindled to about 30, the students evidently preferring to watch the YouTube-hosted videos over showing up in person. In the end, over 20,000 students completed the course, receiving a "statement of accomplishment" but no Stanford credits.

Thrun quit his tenured position at Stanford in 2012 and raised millions of dollars in venture capital to start Udacity, a company dedicated to providing commercial MOOCs. Two other major players have emerged to battle for students in the MOOC arena: Coursera is a commercial start-up founded by Thrun's Stanford colleagues Andrew Ng and Daphne Koller. On the East Coast, edX is the union of the Boston-area Ivies Harvard and MIT, which have pledged a hefty $60 million to fund nonprofit and open-source alternatives to the venture capital–driven Silicon Valley–based MOOC providers.[49]

edX

In late 2011, MIT formed a venture called MITx to explore delivering its courses online. MIT has had a rich history of providing back-of-the-classroom lectures and resources with the OpenCourseWare project, but it was looking for a more comprehensive learning environment that went beyond the unidirectional broadcast model to conceivably exceed, not just equal, the in-person classroom experience.

Buoyed by the initial results from MITx, the MIT team met with Harvard University to discuss collaborating on the development of an open-source platform to provide quality online education and formed the non-profit venture edX. Both schools pledged $60 million in joint investment, made up from institutional support, philanthropy, and grants. Emphasizing quality, MIT's Provost L. Rafael Reif stated that "this is not to be construed as MIT-lite, or Harvard-lite, the content is the same content" as the residential classes.[50] There has been a groundswell of interest from other educational institutions to join Harvard and MIT. UC Berkeley and Georgetown University initially joined with edX, and there are a growing number of prestigious partners, including Boston University, University of Texas, and France's Sorbonne Universités.[51]

Even though edX is a nonprofit venture, it was not designed to be a loss for MIT, Harvard, and the other partnering institutions, so it will begin charging students for activities such as certification and exam proctoring. Both Coursera and edX view their financial relationship with academic partners as a true partnership, where any potential revenues are shared. They are also exploring offering the platform to corporations and non-governmental organizations, such as the International Monetary Fund, and hope to break even in the next three years.[52]

edX has developed the most advanced technologies of all the current developers of MOOCs for the single most difficult element for an online course to scale: assessment. edX instructors have the ability to use innovative tools such as *automatic essay scoring* (AES) software to assess student writing responses at MOOC scales. AES is a technique developed by computational linguists to automatically grade student writing. This is done by comparing the submitted essay with hundreds of other essays on the same topic that have been scored by human scorers, and then return a score that the essay would likely yield when graded by a teacher. More sophisticated AES systems can offer the student precise feedback about what to change to improve the essay.[53]

Perhaps most importantly, because the edX platform is open source, its base operational software is available to anyone to adopt and modify. In addition, edX has made great efforts to open up its internal architecture, making it easy for people to develop and add their own learning and assessment components to the platform through the use of a plug-in architecture called *xblocks*. These components deliver the interactive course content and can be mixed and matched, like virtual Lego bricks, to create customized learning experiences that blend seamlessly with the hundreds of existing xblocks already freely available.[54]

Accreditation

Andrew Kelly, a scholar at the neoconservative think tank the American Enterprise Institute, summed up the accreditation situation nicely, observing that "MOOCs clearly provide new opportunities to learn, but so do public libraries. For degree-seeking students, it is the translation of learning into college credit that endows it with value."[55] In order to facilitate the accreditation process, MOOC providers need to find ways to verify student identities and develop a relationship with other degree-granting institutions willing to accept certified completion of a MOOC as course credit.

Verifying that the student seeking credit is indeed the same person who took the course and its associated assessments is tricky, even in face-to-face classroom settings, where cheating has occurred since colleges began: taking exams for other students, buying prewritten papers, and so on. MOOCs are solving these issues in the same way traditional colleges do—by providing proctored exams in physical locations and using services like TurnItIn.com, where student work is automatically compared with millions of other submitted documents to detect plagiarism.[56]

Getting other institutions to value credits earned in a MOOC is a much thornier problem than verification, and runs up against the very mechanism that most face-to-face institutions use to grant college credit: time in the seat versus competency. Traditional colleges have been resistant to offering competency-based credit, such as simply passing a final exam. What's needed is a trusted third-party mechanism that can independently evaluate a MOOC, determine the academic rigor of the program, and reliably discern the student's level of success with it.

Degree-granting institutions are not completely insensitive to the idea of competency-based accreditation. There are existing models where degree-granting colleges rely on students to pass a trusted examination to receive college credit. The Advanced Placement Program (whose courses are known as AP courses) provides high school curriculum and assessment instruments for 37 courses that many colleges routinely accept as college credit from incoming first-year students. The AP program began after the Second World War, with help from the Ford Foundation, and has been a staple in ambitious students' portfolios for college admission since the nonprofit College Board took it over in 1954.[57]

One downside of AP classes is that they tend to be poster children for a "teaching to the test" pedagogical style. All class activity tends to be focused on memorizing information required to pass the grueling three-hour proctored examinations. For this reason, AP courses often lack the flexibility and liveliness found in most traditionally taught classes. The College Board has another program similar to the AP, the College Level Examination Program (CLEP), which is aimed at providing credit for older students to take advantage of their self-directed learning.[58] There are no requirements in either program for a student to be enrolled in any particular course, just the ability to pass a proctored exam. That said, in order to practically use the AP or CLEP model for accreditation, MOOCs would need to cater their curriculum to focus precisely on what the test was designed to assess. In other words, they would need to teach to the test.

There have been some efforts to use MOOCs as a substitute for introductory and remedial classes, but the results have not always been positive. In 2013, San Jose State University launched a pilot program with Udacity to great fanfare that provided remedial math and statistics classes for college credit using Udacity's MOOCs. In a courageous move, the school stepped back from the program less than year later, saying, "Preliminary findings from the spring semester suggest students in the online Udacity courses, which were developed jointly with San Jose State faculty, do not fare as well as students who attended normal classes."[59]

There has been a recent push for traditional colleges to embrace MOOCs for accreditation. In 2015, the University of Illinois at Urbana-Champaign partnered with Coursera to offer free courses in their MBA program, but students still need to pay tuition ($20,000 for the full degree) to receive their degree.[60] Also in 2015, the Texas University System announced their "Freshman Year for Free" program, where students can earn a full year of college credit through MOOCs offered freely by edX, by passing either the AP or CLEP tests administered by the new nonprofit Modern States Educational Alliance.[61]

MOOC Economics

The economic model for traditional learning institutions is under fire for being too rigid in terms of admission and reliance on physical presence, and for not being cost effective, but MOOCs suffer from the opposite problem. Instead of zealously guarding their content using rigorous admissions standards and high tuitions as a barrier, MOOCs have adopted a low threshold for entry. Like many Silicon Valley technology companies, MOOCs have typically offered their content for free to attract "eyeballs," and then attempted to develop a sustainable business model once they have become popular. This strategy has proven wildly successful for some Internet start-ups such as Google and Facebook, but education is a more constricted marketplace, where revenue approaches such as advertising may not be quite as appropriate or successful.

It's not just the revenue model that is at odds with the peculiar institution of education; the funding model is as well. Investment firms provide venture capital to fund early-stage companies. Venture capital–funded companies receive a certain amount of time to come up with a viable model that can ensure independence on investment capital, but at some point they need to generate revenue. These firms typically invest in a large number of start-ups with the assumption that 90 percent of them will fail, but the 10 percent that thrive will yield a return on investment of at least 300 percent (known in their parlance as a "3-bagger"). This strategy has been extremely successful in the high-tech sector and in large part is responsible for the phenomenal products and companies that have emerged from Silicon Valley. Venture capital firms provide a strong support network to help guide new entrepreneurs, but their model has a darker side.

There is an inherent instability in any "disposable" relationship. The funded companies typically cede a significant amount of control in exchange for the millions of dollars they receive. When the company delivers the kinds of profits that the funders see as significant, that control can be constructive and nurturing. But if the company underperforms or takes longer to deliver, it can find itself among the "walking dead," with just enough capital to stay in business but not enough to grow, closed down completely, or merged with another of the firm's portfolio of funded companies. The venture funding method has worked well in the high-tech sector, but I worry about the effect of this volatility on education. Students have traditionally relied on the longevity and stability of their institutions, and the basic venture capital model may weaken those properties.

A number of business models for MOOCs are currently being explored, and it will take some time to see what eventually emerges as a viable solution. Right now, most companies are offering their courses at no charge to students, but there is no real accreditation to prove their successful completion. Possible revenue sources include charging students for accreditation, ranging from completion certificates, or "nanodegrees," that attest to completion of a specified curriculum, to fully transferrable college credits. Some companies charge fees for additional student support and for proctored exams.[62]

Making Sense of Cloud-Based Media

The delivery of and interaction with media on the Internet have changed the way students access education. Massive amounts of information can be sent instantly at an extremely low marginal cost, and innovators are actively working on ways to use this new medium to its best advantage. What makes this all the more potentially transformative is that the Internet is a dynamic, intelligent medium, mediated by computer technology that is growing exponentially in its sophistication and ability to be a more active partner in the conversation between teacher and learner.

The Khan Academy

Not everyone is enamored with what Salman Khan is doing with his online videos. The educational technology blogger Audrey Watters has characterized the Khan Academy as "Old wine, new bottles, bad pedagogy." The activities of demonstrate, test, practice, test do little to reflect the current constructivist trends in modern classroom education and reinforce the old "drill-and-kill" behaviorist teaching strategies.[63]

David Coffey and John Golden, a pair of University of Michigan professors, posted their own YouTube video, "Mystery Teacher Theatre 2000," which makes fun of mistakes in a Khan lesson on multiplying negative numbers. The pair appears as silhouetted in front of Khan's video and critique it in the style of the old television comedy show *Mystery Science Theatre 3000* (sometimes called MST3K). "It's easy for students to feel successful just repeating a computation they see somebody else do, while not constructing an understanding of the procedures involved," Golden said.[64]

Salman Khan had his own thoughts as to why some educators might not be so supportive, saying, "It'd piss me off, too, if I had been teaching for 30 years and suddenly this ex-hedge-fund guy is hailed as the world's teacher." Some of Khan's early videos had errors, and his explanations often lacked proper reasoning, or focused on the mechanics of solving the problem rather than providing insight at the conceptual level.[65] To their credit, the Khan Academy has brought in a number of content specialists to address these accuracy and pedagogy issues. The current work in the academy's computer science and programming curriculum is showing some innovative and constructivist techniques in this notoriously difficult-to-teach discipline.

MOOCs

After teaching a MOOC on artificial intelligence to over 100,000 students, Udacity's founder Sebastian Thrun remarked, in an homage to the film *The Matrix*, "I feel like there's a red pill and a blue pill, and you can take the blue pill and go back to your classroom and lecture to your 20 students. But I've taken the red pill, and I've seen Wonderland."[66] But not all educators are willing to swallow the red pill. The faculty at Amherst College voted to decline an invitation by edX to join as a partner. Amherst law professor Adam Sitze asked rhetorically, "What makes us think, educationally, that MOOCs are the form of online learning that we should be experimenting with? On what basis? On what grounds? 2012 was the year of the MOOCs. 2013 will be the year of buyer's regret."[67] The excitement and apprehension stirred up in 2012 has settled down in recent days, and slow but steady progress is being made in how to solve some of the problems posed by this new form of educational content delivery, and where its place in higher education might be.

In the end, what is most enticing about MOOCs is their ability to scale instruction so dramatically. If a typical college classroom might have a 1:25 instructor-to-student ratio, a MOOC can potentially have a 1:100,000 instructor-to-student ratio. This kind of leverage is appealing in an era where the costs of higher education are spiraling out of control. But MOOCs now can only do this using a simulacrum of the often-ineffective traditional lecture experience. Instead of employing more constructivist strategies that encourage higher-order thinking skills, MOOCs instruct using a content-driven, more didactic pedagogical approach: a single instructor delivering content to large numbers of students who are constantly assessed with multiple-choice questions, rather than engaging in a dialogue with others.[68]

It will take some time for the Internet-based instructional technologies to evolve beyond the mere recapitulation of previous educational strategies—echoing what Marshall McLuhan first observed about media in general—but there has much interest in pushing the boundaries. Chapter 5 examines educators' early use of more immersive technologies that try to situate the learner in a rich, multisensory environment.

5

Immersive Media

If I had a world of my own, everything would be nonsense.
Nothing would be what it is,
because everything would be what it isn't.

Alice

As computers became more capable of delivering realistic multimedia experiences, a new teaching tool emerged: the immersive learning environment. Computer and video games for entertainment were the first to take advantage of this new realism, where the player feels completely immersed in a virtual world represented on the screen. Educators soon began developing more immersive instructional applications, hoping to harness this sticky phenomenon toward more fruitful ends. Immersive learning tools take advantage of the personal computer's ability to deliver more lifelike renderings of scenes and offer constructivist approaches to explore the synthetic portions of them.

In many ways, this work is the logical extension of geographic navigation of a three-dimensional space in two dimensions, hinted at by Berkeley's simulations of urban landscapes using the Modelscope camera, and the Massachusetts Institute of Technology's use of the videodisc to virtually traverse the town of Aspen. Hypermedia, HyperCard, multimedia, and the Internet itself provided new tools to navigate the decidedly more abstract information-based world of using spatial metaphors, but the ability to realistically render realistic-looking 3D scenes made more immersive experiences possible.

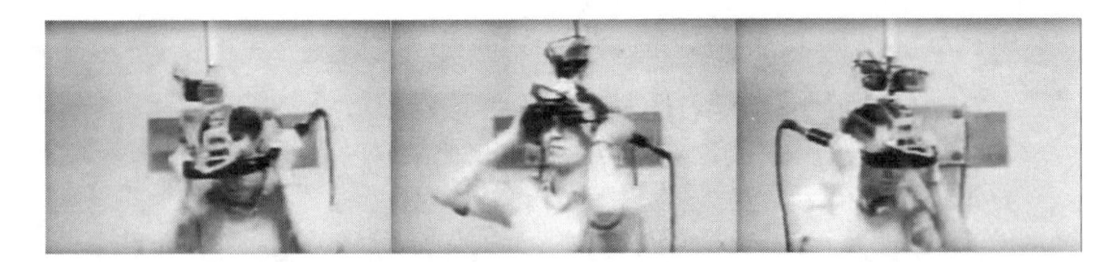

Figure 5.1. Ivan Sutherland's head-mounted virtual reality display.

Computer graphics pioneer Ivan Sutherland created some of the first immersive 3D environments in the late 1960s. These were expensive affairs, with small monitors mounted inside of helmets (fig. 5.1) that sensed where the wearer was looking, and they quickly rendered a new 3D scene in real time to match that position, so the wearer felt as if she were actually in the environment.[1] This early research launched the field known as *virtual reality* (or VR) and was funded by the military to train soldiers and airplane pilots using flight simulators. Today, simpler, less expensive, less intrusive devices offer VR experiences, including the Oculus Rift and even a free cardboard kit from Google, which both use smartphones as the image source.[2]

These immersive worlds are familiar to video game players today, and they can provide an ultrarealistic environment where the participant can roam virtually through the setting using the mouse or keyboard. Inside this synthetic environment, participants are represented by graphical representations known as *avatars*.* Most simulations provide ways for users to customize the look of their avatars and even elaborately clothe them. In some simulations and games the avatar is visible to the participant, as if being filmed by a cameraman slightly behind them, while others (known as first-person shooters) simply show the participant as if he is seeing things through his avatar's eyes.

Chris Dede, an early immersive-world education researcher, defined the immersive experience as having "the subjective impression that one is participating in a comprehensive, realistic experience. Immersion in a digital experience involves the willing suspension of disbelief, and the design of immersive learning experiences that induce this disbelief draws on sensory,

* The term "avatar" is derived from Hinduism and refers to a representation of deity in a terrestrial form.

actional, and symbolic factors."[3] The actional portion of his definition calls for the learner to participate actively in this new world and even initiate actions that might be physically impossible in the real world, such as becoming a bird and flying, to better understand that impractical experience.

Immersive environments can also enhance learning by offering multiple perspectives on a situation. A student can shift from her own frame of reference to someone else's, which can provide valuable context for understanding complex concepts. Like the *Jasper Woodbury* videos, immersive environments provide a strong contextual base for situated learning and anchored instructional techniques—especially when combined with expert modeling and mentoring. Finally, immersive environments can help facilitate the all-important transfer of knowledge into the real world, as successfully evidenced in many flight and surgical simulations.[4]

These simulations allow instructors to fashion an idealized environment that helps students learn in a new world that is stripped of messy reality. The real world always has extraneous things that can distract students from the task at hand, and these simplified learning environments help them focus on issues germane to learning some specific concept or skill. Immersive worlds are not useful in all situations, however, particularly those that involve deep reflection, the analysis of longer documents, or where sustained discussion is necessary.[5]

Immersive Learning

Immersive learning environments have been strongly influenced by the Swiss developmental psychologist Jean Piaget's theory of *constructivism*, and Seymour Papert's later work on *constructionism* with the Logo programming environment. Piaget had strong ideas on how children viewed the world they inhabited. He saw them not as merely smaller versions of adults, but as distinct beings with age-related views of their environments that are fundamentally different. For Piaget, successful education began with an understanding that children had a set of evolving theories based on direct experience that related to previous experiences and understandings. This, in effect, was how children created meaning. In order to expand upon current understanding, children need to discover for themselves the "error of their ways" and make connections based on their own experience rather than learning by being told directly.[6] Piaget believed that children are "builders of their own intellectual structures" without any need for formal instruction, that education came from the bottom–up (as opposed to top–down), and that learning occurs without a formal curriculum and without deliberate teaching.[7]

Seymour Papert was a student of Piaget's whose ideas diverged from his in some subtle but important ways. While both men thought that all new knowledge was self-constructed, Piaget held that children learned higher-level formal and abstract ideas from concrete interactions, while Papert believed the opposite. Papert's constructionism posits that the concrete instance and the abstract idea it represents are part of the same continuum and need to be considered together, saying "becoming one with the object under study is a key to learning."[8]

This view of learning contrasts sharply with the reductionist methods used in much of the earlier educational technology attempts, where larger ideas were chopped up into smaller, more digestible chunks, and the larger abstract or gestalt emerged from the pieces. Papert created simplified environments he called *microworlds*, where instead of leading the student to the answer in small steps, he set up a simplified environment that was constrained for exploring some specific problem, in which the student would construct his own solution to the problem.[9] In constructionism, the teacher's role is to define the problem or concept to be learned, construct an environment that will facilitate solving the problem through exploration and inquiry, and provide gentle support of the student's exploration of that environment.

Immersive environments offer a new way to create these finely focused learning microworlds. Educators found a number of different ways to create an immersive learning experience, ranging from simulations of specific situations—not unlike technology-enhanced version of the *Jasper Woodbury* adventures—educationally driven games, and explorable virtual worlds. There is considerable overlap between the categories, but Koreen Pagano's taxonomy (fig. 5.2) is a useful way to categorize them.[10]

- *Contextual decision-making experiences* create a virtual environment that provides a microworld for learners to explore in a relatively structured manner. They are most often rendered as dynamic simulations of that world, highlighting salient contextual elements and providing scaffolding that gently guides learners through that environment toward some kind of goal.

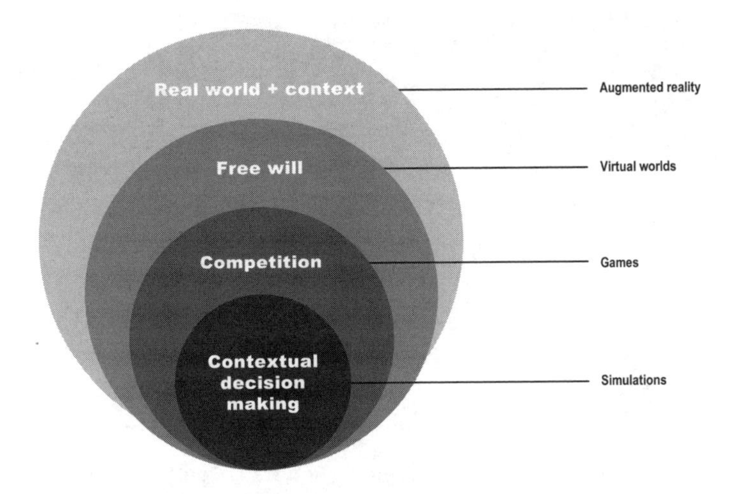

Figure 5.2. Immersive learning categories. Adapted from K. Pagano, *Immersive Learning: Design for Authentic Practice* (Alexandria, VA: ASTD Press, 2013)

- *Competitive experiences* create an environment where the learner is in competition with herself, with others, or with some set performance standard to complete a predefined goal. These are often in the form of games. Successful competitive learning experiences provide rapid feedback as to the learner's progress toward the goal.
- *Free-will experiences* are like simulations, but with less explicit structure and scaffolding. They may or may not have defined goals and are typified by virtual worlds.
- *Real world with additional context experiences* blend the virtual rendered environment with elements from the real world, creating what is called *augmented reality*. These can be GPS-fueled navigation of virtual scenes based on real-world coordinates, or media overlays of images, annotations, and graphic on top of video captured in real time.

Degrees of Immersion

In order for an immersive environment to be effective, and the learner truly immersed within it, a certain amount of faith in that environment is needed. Such *suspension of disbelief* is critical to the enjoyment

Figure 5.3. The continuum of representative detail. Courtesy of Scott McCloud

of all types of human-designed experiences, from books to motion pictures to video games. The poet who inspired Hypertext pioneer Ted Nelson, Samuel Taylor Coleridge, coined the phrase in 1817: "my endeavours should be directed to persons and characters supernatural, or at least romantic, yet so as to transfer from our inward nature a human interest and a semblance of truth sufficient to procure for these shadows of imagination that willing suspension of disbelief for the moment, which constitutes poetic faith."[11]

There are degrees of immersion, and different situations require varying levels of that faith. The cartoonist and educator Scott McCloud wonderfully illustrates (fig. 5.3) this point by showing the degree to which people can generalize the actions of the figure being drawn on multiple continua: complex to simple, realistic to iconic, objective to subjective, and specific to universal. The simpler (or more abstract) the presentation, the more the viewer can see the drawing as more broadly generalizable, and the more detailed the representation, the more likely the drawing will be interpreted as a particular item or phenomenon.[12]

The psychologist Mihaly Csikszentmihalyi (pronounced "Me-high Sent-me-high") has studied a phenomenon he called optimal experience, or *flow*, when the person becomes so involved with whatever activity he is doing that all other things fade into the background, resulting in a highly pleasurable state of focused attention. He described it as "what the sailor holding a tight course feels when the wind whips through her hair, when the boat lunges through the waves like a colt—sails, hull, wind, and the sea humming in a harmony that vibrates in the sailor's veins."[13] Athletes experience flow when they are "in the zone," and the novelist David Foster Wallace described his exhilaration while writing fiction under a state of flow as when "I can't feel my ass in the chair."[14]

Some immersive experiences such as video games routinely engage participants in a state of flow, but this desired outcome is rarer in educational applications, as much as their developers dream for it. And the degree of immersion may not correlate to how realistically the experience is rendered. Players who engage with online text-only interactive games routinely suspend their disbelief more than people who don head-mounted 3D displays that react to every movement of their heads.

But there can also be too much of a good thing. Designers need to be aware of just how realistically the depictions of people in their immersive experiences are rendered. If the depiction of people becomes too real, the participant may get a vague feeling of unease known as the *uncanny valley*. Since the first industrial robot was made in 1959, robotics developers sought to make their robots evermore realistic in the belief that people would be more likely to accept robots that seemed human. This proved to be true, especially for children, but only up to a point.

Masahiro Mori, a robotics professor at Tokyo Institute of Technology, drew a hypothetical graph[15] in 1970 that plotted people's positive reactions versus how realistic a depiction a human likeness becomes (fig. 5.4). He predicted a dip in positive affect just before 100 percent realism, and called this drop in the curve the uncanny valley (*bukimi no tani* in Japanese). Up to that point in the realism continuum, people would view the robot as a purely nonhuman device; after it, they would view it as human, but the space in between would give people an uncomfortable uneasy feeling of dread. Mori recommended that designers aim for the point just before the valley dips, rather than 100 percent realistic, because all the points in between will cause uneasiness among viewers.[16]

Trouble in River City

In 2002, a group of Harvard researchers led by Chris Dede wondered whether an immersive environment could be used to teach the process

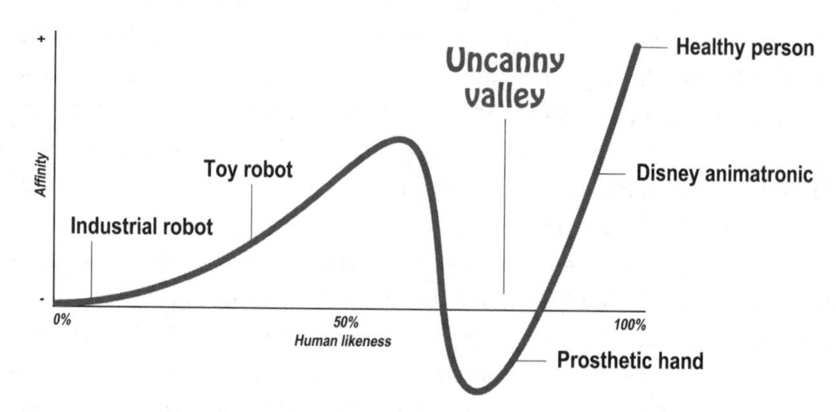

Figure 5.4. The uncanny valley. Adapted from M. Mori, "The Uncanny Valley," *Energy* 7, no. 4 (2012)

of scientific inquiry in a manner that was more engaging than traditional classroom instruction. Using funding from the National Science Foundation (NSF), they designed a multiuser virtual environment (MUVE) that immersed learners in a participatory historic 3D simulation, and ran a preliminary study with a small group of middle school students. A MUVE is a networked simulator where multiple participants are placed in an interactive 3D simulation and can "roam" through that synthetic world and interact with one another as if the world were real.

The goal of the pilot study for the researchers was to see which factors may be important to foster the critical thinking skills needed for scientific inquiry, similar to those needed by students to participate in a science fair project. The curriculum was provided in both Spanish and English and aligned with state educational standards in biology and ecology. The immediate task for the students was to better understand the scientific method and then to apply it in other situations.

The 3D simulation was based on a small nineteenth-century industrial city with a river running through it, called, oddly enough, *River City*. The virtual town was historically accurate, replete with digital objects from the Smithsonian Institution, and had a hospital, hotel, university, shopping area, and a number of residential neighborhoods (fig. 5.5). As in *The Adventures of Jasper Woodbury*, the students are placed in a data-rich environment and asked to solve a problem. The people of River City are getting

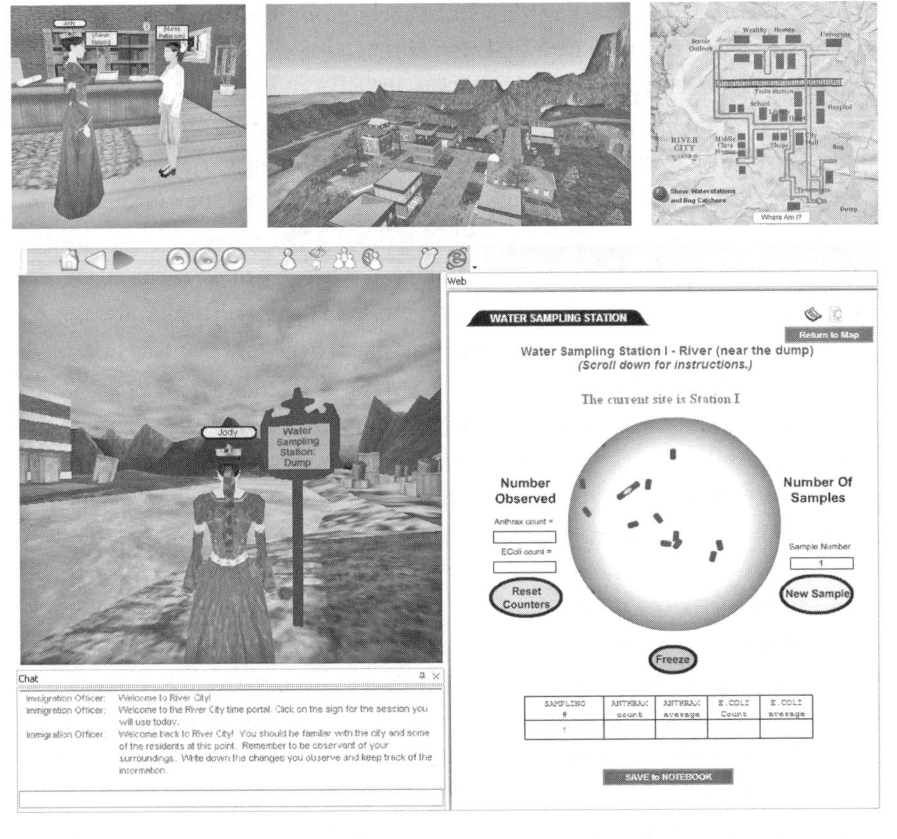

Figure 5.5. Screens shots from the *River City* project. Courtesy of Chris Dede

sick and the mayor needs to know why. The residents are coming down with three different diseases, each of which have a number of causal factors, such as terrain, water runoff, water quality, mosquitos, the economic status of the neighborhood, and so on. The goal is to decide which aspect(s) of the city should be altered to reduce illness among the city's residents.

To help the mayor, the students need to do the same basic tasks in devising a traditional scientific research plan: they need to develop an idea of what might be causing the illness (a hypothesis), identify the parameters that could be making things better or worse (the independent variables), and find measurable data to determine whether changing those parameters would have any effect (the dependent variables). Students work in teams of three to go "back in time" and gather longitudinal data from four consecutive "visits" to the virtual city, from October 1878 to July 1879, followed by two face-to-face classroom sessions.

One unique feature distinguishing *River City* from *Jasper Woodbury* is that the students can change the various independent variables and then see what results they might have. On subsequent visits, they measure the effects those changes may have had in reducing the spread of the illness, and ultimately write a recommendation to River City's mayor. The students use built-in data collection and analysis tools, such as a virtual microscope, and iteratively test and refine their hypotheses based on the data they have collected and analyzed.

Learning in *River City* is an inherently social process. While in the simulated environment, students interact with computer-based agents who play the roles of River City residents, and students can also eavesdrop on the private conversations these nonhuman residents have with one another to gather more data. Students in the simulation can interact with one another and with the nonhuman residents using a text-driven chat box. As a nice alternative to the multiple-choice test for formative assessment, a computer-based agent playing the role of a newspaper reporter interviews the students for information one or more times per visit about a story he is writing about the illness situation, which helps teachers assess what the students know.

One year after the 2002 pilot study, the NSF funded a more ambitious trial of *River City*, with over 250 teachers and 15,000 students ultimately using the simulation. As a part of that study, the researchers ran an experiment that compared students using the 3D simulation environment with students taught using the same curriculum but with paper-based methods. Students who used the simulation software had scores on a biology assessment that were 16 percent higher than the control group using more traditional materials.[17]

The researchers were careful to include feedback from the classroom teachers involved in the NSF study and iteratively incorporated changes to the implementation process. They also developed extensive support materials for teachers using *River City* in their classrooms. Like many university grant–funded projects, Harvard no longer directly supports the *River City* simulation, but a commercial company still provides access to the software for about $10 per student.[18]

Second Life

Second Life is perhaps the most ambitious online immersive environment attempted to date, with well over a million users at its peak in 2007. It is a commercial enterprise from Linden Lab, which funds itself by selling and renting virtual "land" that people purchase and then improve upon, just as in the real world. It attempts to create a simulacrum of the real world with a relatively realistic and immersive three-dimensional shared environment, replete with a functioning economy; a robust real estate market where people buy and sell virtual land; and a flourishing commerce in objects that people can make, buy, and sell to furnish their properties and even clothe themselves.[19]

The company provides the raw land, but it's up to the residents to construct things like buildings on it. Linden Lab has created a digital currency called Linden dollars, for use "in world" (that is, when participating in the *Second Life* environment) that can be used to buy and sell digital objects, clothes, and the like. Linden dollars can be exchanged with real dollars, with one US dollar being worth about 260 virtual dollars. To date, over $60 million of Linden dollars have been bought.[20]

This is what makes *Second Life* so different from other virtual worlds; it begins as a blank slate and is entirely user created. The software has fairly robust built-in 3D drawing tools for residents to create these objects with and sell to others. There is also a powerful scripting language that allows for complex animations and interactions with users, allowing for potentially rich experiences. *Second Life* has attracted large numbers of educators who have used it to hold classes, provide spaces for students to meet with one another, and offer more elaborate teaching experiences similar to the *River City* project.

Second Life was the brainchild of Philip Rosedale (fig. 5.6), the son of a US Navy pilot who grew up in a semirural tidewater town in St. Mary's County, Maryland. He became interested in programming his own version of Stephen Wolfram's cellular automata simulations and was amazed that so much complexity could emerge from a few simple rules. When the first immersive PC games like *Doom*—one of the first immersive multiplayer video games, created by legendary game developer John Carmack in 1992[21]—began to appear, he wondered, "How can I do this and how can I be in the machine? Not be in the machine and play *Doom*, but be in the machine and build things?"[22]

Figure 5.6. Linden Lab founder Philip Rosedale, 2009. Courtesy of Roland Legrand

After college, Rosedale moved to San Francisco to start a small Internet-based company that compressed video streams to better flow over the slow dial-up connections of the time. His company attracted the attention of Rob Glaser's streaming media company, RealNetworks, which acquired his company in 1995. Rosedale served as RealNetworks' chief technology officer until leaving in 1999, yet another smart, ambitious technology millionaire looking for his next big idea.

After trying a number of ideas in their downtown San Francisco warehouse office on Linden Street (the origin of Linden Lab's name), Rosedale and his colleagues found an idea that delivered on his wish to be in the machine and build things. They were inspired by the edgy Burning Man festival, held every year at an empty Nevada desert where people massed and constructed a community complete with an informal economy. Originally called *Linden World*, the goal was to provide a blank space where people could construct their own community and interact online. They later changed the name to *Second Life*, a riff on Milton Bradley's popular board game *The Game of Life*.

Rosedale's time at RealNetworks was a good foundation for crafting how *Second Life* was delivered over the Internet. Like audio and video files,

3D models are large and take a long time to download. Rather than load them upfront, *Second Life* streams the models in the same way RealNetworks streams media. Rich 3D scenes appear on an "as-needed" basis, minimizing time needed to begin viewing and resulting in a better user experience.[23]

The largest unit of land available in *Second Life* is an entire island, which can be purchased directly from Linden Lab, but there are a number of real estate agents who will sell "used" areas of land and also rent portions of other people's islands. An island costs around $1,000 to purchase, and Linden Lab offers a 50 percent discount to educators. Beyond the cost of buying land, Linden Lab charges island owners a fee of $295 per month.[24]

The purchase of educational islands peaked in 2007, when a poor business decision by Linden Labs eliminated the education discount, causing bad feelings between the company and potential education customers. The discount was eventually reinstated, but schools never really got over the issue and have stayed away from purchasing islands ever since.[25]

In World with Wendy

Had I taken chemistry in college, Wendy Keeney-Kennicutt is the kind of instructor I would have liked to have. She has been teaching first-year chemistry classes at Texas A&M since 1984 and has garnered an impressive number of teaching awards along the way. Chemistry as a discipline is not typically kind to beginning students, and it often serves as a "gatekeeper" course to thin out the pack of first-year students aspiring to become doctors. It is an abstract and difficult subject that requires students to solve a lot of tedious problems in order to be successful. Chemistry professors are typically reviewed harshly on RateMyProfessors.com, a site where students publicly rank and comment on their instructors, but comments such as "Best chemistry professor I have ever had!" and "She just wants her students to be successful" were common in reviews of Keeney-Kennicutt's teaching.[26]

Wendy Keeney-Kennicutt (fig. 5.7) grew up in Upstate New York wanting to be a doctor. She went to school at Queen's University in Canada because it was cheaper than most colleges in the United States. Instead of being weeded out by chemistry during her undergraduate studies, after serving a year in the Volunteers in Service to America (VISTA) program,

Figure 5.7. Wendy Keeney-Kennicutt. Courtesy of *The Battalion*

she went on to get a master's degree in chemistry and ultimately received her PhD in chemical oceanography from Texas A&M in 1982.

After a few years of teaching first-year chemistry classes at Texas A&M, she found the experience of lecturing boring, and not resonating well with her students, saying, "Chemistry hasn't changed in 40 years, but kids have."[27] In 2009, the university began experimenting with *Second Life* as an instructional vehicle, initially buying three islands for instructors to use in their teaching (they later bought over a dozen).[28] University officials recommended that Keeney-Kennicutt look into it for her introductory chemistry classes.

After trying it out, she became convinced that an immersive 3D environment like *Second Life* would be useful for teaching chemistry. Practicing chemists need to be able to visualize the arrangement of atoms that make up a molecule in 3D space, and mentally and fluently transform molecules between the 2D and 3D worlds. In 2011, she ran a small study using *Second Life* to teach first-year chemistry students a concept that required these kinds of 2D and 3D abilities. She was able to script much of the interaction in the elements herself, and, using Linden dollars, she purchased models and lab equipment from other universities and got free modules from the

American Chemical Society, which had invested heavily in building modules such as an interactive periodic table of elements. In *Second Life*, she could show PowerPoint presentations, go on virtual field trips, and have the students interact with 3D models of molecules, answering questions about them while in world. The control group performed similar activities using static images of the molecules instead of *Second Life*.

She held virtual office hours inside *Second Life*, as well as traditional face-to-face meetings, and students voted with their feet (and their mouse) as to which venue they chose for help. Students who were comfortable with *Second Life* came to visit her in world, where she talked using a headset and they typically chatted using the keyboard. Becoming comfortable in *Second Life* requires a steep learning curve, and the students needed practice navigating the sometimes-awkward 3D space. The students who had experience in virtual worlds through immersive games managed the navigation well, but they still faced the typical problems of implementing technology in an educational environment: bandwidth, security firewalls, and getting their microphones to work properly.

The study of approximately 600 students yielded no significant difference between the control group and the students that used the virtual environment, but even so Keeney-Kennicutt continued to use *Second Life* in teaching introductory chemistry, adopting the motto of "do no harm" as the criterion of success during this stage. Some students thought the dimensionality helped them better understand the basic structures than traditional 2D representations, but others found the overhead involved in navigating the 3D environment more trouble than it was worth.[29]

Buoyed by the pilot study and interested to see what potential this new way of teaching might have using a more interactive laboratory experience, she partnered with Kurt Winkelmann, a chemistry professor from the Florida Institute of Technology, to ask the NSF to fund a more robust and interactive environment study of teaching chemistry to thousands of students, beginning in 2014.[30] The researchers used the internal 3D drawing tools in *Second Life* to craft virtual laboratories that looked exactly like physical chemistry labs in the expensive Texas A&M chemistry department (fig. 5.8). These stations looked realistic, but one could not actually do chemistry experiments there. They used $30,000 from the grant to hire a pair of programmers to create several levels above the workstation where students could perform experiments using the same steps that students in

Figure 5.8. The real and virtual chemistry labs at Texas A&M. Courtesy of Wendy Keeney-Kennicutt

the physical labs used in scripting language built into *Second Life*. Keeney-Kennicutt said, "No other tool could mimic a real lab the way *Second Life* could."[31]

Professor Keeney-Kennicutt invited me to join her in world at the virtual Texas A&M chemistry lab to try out some of the virtual experiments under her guidance. I signed on to *Second Life* and chose an avatar to represent myself. (It turned out to be younger and slimmer than I am in real life, but, honestly, the choices were limited.) Keeney-Kennicutt was similarly represented and she met me in Chemistry World, one the university's virtual islands, where we walked into a classroom (fig. 5.9A) in the chemistry building.

She got dressed in a white lab coat, I assumed to protect her avatar's sensitive clothing, and we entered the lab itself. We then went upstairs set to do the actual experiment: to find the molar mass of butane gas (fig. 5.9B). Using choices from on-screen menus, we first weighed a butane cigarette lighter using the digital scale tool. We also measured the atmospheric pressure and room temperature. We then filled a basin with water from a spigot and placed a graduated cylinder upside down in the water. We attached a small hose to the lighter, which was placed on the other end of the hose under the cylinder, and let gas collect until a certain amount of the water was displaced (fig. 5.9C). Then we weighed the lighter again and had the all data needed to solve the problem.[32]

The experience replicated how one would do the experiment in a physical lab, less the potentially wet clothing. I must confess that navigating the three-dimensional space was frustrating at times. I kept getting lost and

Figure 5.9. Scenes from the virtual chemistry lab at Texas A&M

stuck in corners facing the wall, much like the hiker in the *Blair Witch* film, but my host was a patient guide. At one point, I was inexplicably sent onto the roof, and she had to teleport me back down to the lab. I could definitely see how some students might find the experience tiresome, but at the same time, my progress through the experiment was well scaffolded by menus that made the next steps clear.

Virtual labs could provide a reasonable alternative to the expensive real-world laboratory experience for distance learning and schools that can't afford physical setups. With some active computer-based feedback guiding the student away from making wrong decisions early on, virtual labs could potentially be as good as or better than their real-world counterparts.

Augmented Reality

Just a year after publishing *The Wonderful Wizard of Oz* in 1900, L. Frank Baum wrote a novel dedicated to his 15-year-old son Robert, about a boy who accidently conjures a demon after experimenting with an electric apparatus. This Demon of Electricity proceeded to give the boy a series of gifts, including this one: "I give you the Character Marker. It consists of this pair of spectacles. While you wear them, every one you meet will be marked upon the forehead with a letter indicating his or her character. The good will bear the letter 'G,' the evil the letter 'E.' The wise will be marked with a 'W' and the foolish with an 'F.' The kind will show a 'K' upon their foreheads and the cruel a letter 'C.' Thus you may determine by a single look the true natures of all those you encounter."[33]

The boy ultimately decides that mankind is not quite prepared for the apparition's gifts and tells him to wait until the world is ready for them. Nearly a century later, it appears that the world is ready for at least one of the demon's gifts. Baum's Character Marker neatly describes the emerging field of augmented reality, or AR, and a number of corporate spirits are trying to deliver the demon's gift using computer technology. Where the virtual reality systems of simulators and *Second Life* try to completely envelop the user within a virtual environment, augmented reality systems, like the Character Marker, embellish a view of the real world with synthetic 3D objects that look as though they are embedded within it.

These new systems use a combination of GPS, the device's internal compass, and image recognition software to decode the geometry of the overall scene. The AR user looks at the real world through a video camera in a smartphone, tablet, or special goggles. When that orientation has been determined, 3D objects and annotations that seem to fit perfectly can be inserted into the scene, and move as the user's camera changes the view. AR is a form of mixed reality, where the synthetic world blends seamlessly into the real world, and those synthetic elements help contextualize the scene in some manner.

Augmented Reality Goggles

The search giant Google has been an active participant in providing hardware for both virtual and augmented reality systems. In 2013, they released a head-mounted display called Google Glass, which used an internal compass, GPS, and image recognition to interact with the real world

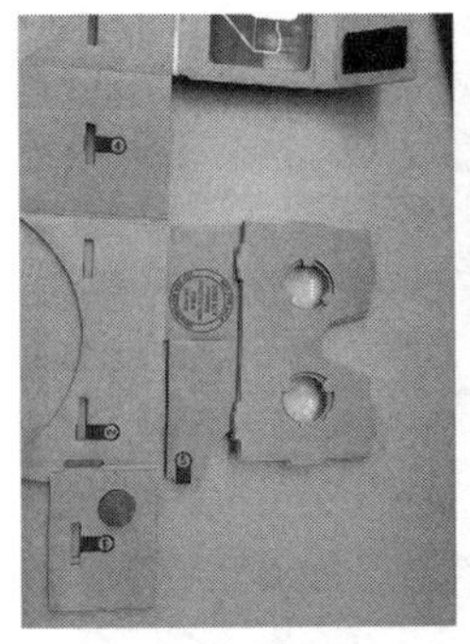
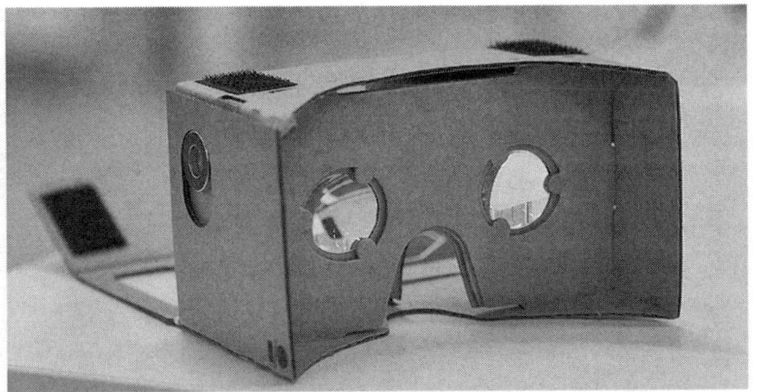

Figure 5.10. A Google Cardboard VR display. Courtesy of Ofthree

and overlay interactive objects. The company has since stopped selling the expensive ($1,500) device, but it prompted much excitement about the consumer possibilities for AR, as well as a flurry of discussion about the privacy issues of recording people without their permission.

Google also threw a bone to the low end of the AR market with a clever device, Google Cardboard, that could turn smartphones into virtual-reality headsets for just pennies. They announced this device at Google I/O, an annual event for software developers, hosted by Google to drum up enthusiasm for its new high-tech initiatives. Like attendees at the prestigious Academy or Emmy awards ceremonies, Google I/O participants have come to expect to go home with a gift bag, filled with expensive perks for attending. Expensive gifts like an Android phone or Chromebook laptop were not uncommon. Attendees at the 2014 Google Developer's conference were surprised to find a folded-up piece of cardboard in their goody bags. When assembled, it looked like an old View-Master stereoscopic viewer toy, but it held a smartphone instead of a round disc of images in front of the viewer's face (fig. 5.10).

When the smartphone was loaded with specially designed apps, the contraption became a surprisingly robust AR tool. Just as Google intended, it became a catalyst for developers to create applications to expand on AR's potential. There are now hundreds of apps, ranging from games and guided tours of the constellations to a live performance of Sir Paul McCartney's "Live and Let Die" in 360 degrees. It has helped gain awareness for other slightly more sophisticated but affordable VR devices like the more recent headset, the Oculus Rift.[34]

Figure 5.11. Augmented reality with the Mars Curiosity rover. Courtesy of NASA's Jet Propulsion Laboratory

Microsoft has launched a large internal effort to bring AR into the mainstream with its HoloLens project, which is a full VR/AR computer system bundled into a headset. HoloLens presents high-resolution holograms (think the "Help me, Obi-Wan Kenobi, you're my only hope" scene from *Star Wars*[35]), seamlessly embedded into the real-life scene the user is looking at.[36] They are not traditional holograms, of course, which use photographic film to capture a light field around an object, and the use of the name has been the subject of much criticism from the research community.[37]

Microsoft is working with Case Western Reserve University to use the HoloLens to teach human anatomy to medical students. Instead of relying solely on cadavers and flat illustrations in textbooks, medical students can interactively explore the body's many intricacies, collaboratively and from multiple angles. The various body layers can be made translucent so that students can see details contextualized within a whole working anatomical system. One Case radiology professor explained, "You see it truly in 3D. You can take parts in and out. You can turn it around. You can see the blood pumping—the entire system."[38]

Microsoft has also partnered with NASA's Jet Propulsion Laboratory in a fascinating use of AR to interactively explore the surface of Mars using data from the Curiosity rover (fig. 5.11). Since landing on Mars in 2012, Curiosity's rich imagery and data have been invaluable to scientists hoping to better understand the red planet, where they previously had a difficult time remotely orienting themselves using two-dimensional images.

The Jet Propulsion Laboratory used the data and images to reconstruct a 3D view of the planet's surface and allowed scientists to roam through this virtual rendition of the planet using Microsoft's HoloLens goggles. They could "walk out" on the Mars surface, add annotations, and collaborate with other team members, as if they were actually on the distant planet, and better understand the topography and identify places for the physical rover to examine.[39]

Augmented Reality on Smartphones and Tablets

The fact that smartphones and tablets have increasingly powerful microprocessors and memory capacities that rival desktop computers, and have high-resolution video cameras so closely embedded with their equally high-resolution displays, has enabled a new class of AR tools. Combined with simple and compelling applications, these devices can often provide as rich and immersive experience as the more intrusive head-mounted goggles, but with a lot less cost and trouble. Because iPads and other tablets are becoming more commonplace in the classroom, a number of education-oriented apps are emerging that have the potential to make learning experiences that are as engaging as they are effective.

Drew Minock was a third-grade teacher in Bloomfield Hills, Michigan, with a keen affinity for new technology. He was one of those people who "waited in line to get the first iPhone," and had a strong desire to use the new technology in his teaching. His co-teacher, Brad Waid, was like-mindedly bullish on the idea, and together they actively sought out technology that might enhance their classroom teaching. While at an educational conference in 2013, Minock saw a presentation of AR applications for tablets and smartphones, and he was taken with the potential for his classroom, thinking, "It was like a magic trick. My jaw hit the floor." [40]

Bloomfield Hills is an affluent town about 25 miles north of Detroit. Each of their school's classrooms had five Apple iPads, so the two co-teachers began to experiment with using AR applications during their classes. Minock and Waid tried one to teach reading and word-recognition skills by overlaying compelling 3D animals on the iPad's screen when the camera was pointed at printed flash cards.[41] Another app helped bring

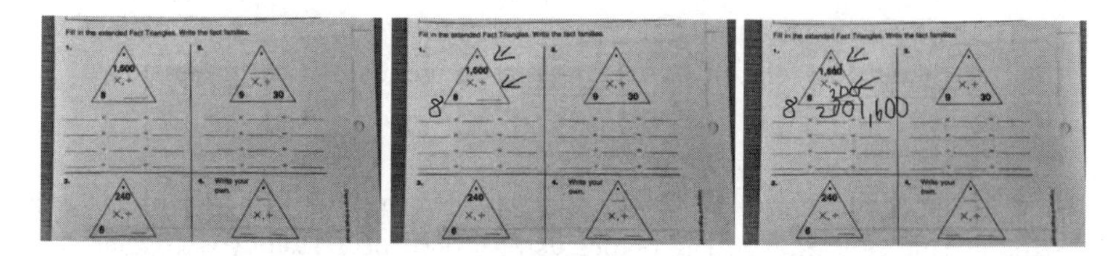

Figure 5.12. Overlay on a student's worksheet. Courtesy of Drew Minock

human anatomy alive by pointing at flat sheets of paper, known as trigger sheets, which the app recognizes and then overlays animated 3D anatomical parts, such as a beating heart.[42] The two teachers found that their students responded well to the new technology.

Because most of their students had AR-capable tablets or smartphones available at home, Waid and Minock took advantage of their school's relative affluence and created one of the more compelling examples of AR in education I've heard about thus far. Like many elementary teachers, they often assign math problems for homework using printed worksheets that the students solve, fill in, and return the next day. Just as flipped-classroom pioneers Jonathan Bergmann and Aaron Sams noticed that kids often got stuck trying to complete worksheets, Waid and Minock, too, looked toward technology to provide immediate help to keep the students on task. Instead of simply providing a video to support the activity, Waid and Minock found a way to use their students' tablets to deliver that video precisely within the context of the worksheet.

Minock recorded a video that showed him working through one of the assigned worksheet problems with a pencil while narrating the process (fig. 5.12), in a way similar to what Salman Khan does in his YouTube videos. If a student became stuck, they could point their tablet over the problem and see Minock's freehand writing overlaid atop their own worksheet, as if he were sitting beside the student explaining it in person. The result is compelling for the learners; Minock adds, "It connects you with the student."[43]

The two teachers energetically proselytized their work using AR with an active speaking schedule and informative blog[44] and have since joined the AR helmet manufacturer DAQRI as AR educational evangelists. The company has shown a strong commitment to education, providing a number of free AR apps and resources to K–12 classroom teaching. One example of these apps is Elements 4D[45] for the iPad, iPhone, and Android tablets,

Figure 5.13. Elements 4D AR app. Courtesy of DAQRI

which uses physical cubes that kids manipulate with their hands to explore how atoms combine to make molecules. Students are supplied with a pair of large dice-like cubes, with beautiful drawings of elements on each face that represent atoms, such as sodium or chlorine. Pointing the camera to a cube's face brings up information about that atom on the screen (fig. 5.13). When the student moves two atoms that can combine with one another, the screen shows information about the chemical reaction and the new molecule that is formed by their union.

Making Sense of Immersive Media

Virtual reality environments can immerse students in microworlds where they can authentically solve problems in a semirealistic environment that creates situations for true situated learning experiences. VR has been wildly successful in military and flight training because it allows learners to safely experience potentially dangerous situations where they can make mistakes out of harm's way. Harvard's *River City* simulations successfully guided students to better understand the abstract world of the scientific method in a manner that students liked better than simply reading about it.

Schools experimented with using immersive multiplayer environments such as *Second Life* to create virtual spaces for instruction, with varying degrees of success. These tools proved to be somewhat awkward instructional spaces, having been designed more for social entertainment than education and needing to compete in a commercial environment. Texas A&M's virtual chemistry labs in *Second Life* can be as effective learning spaces as physical labs for some students, and they provide a real option for learners in remote locations without access to real-life chemistry labs.

I am more sanguine about the potential of AR systems in education, because they can bring instructional context to real-world environments. This mixed reality can be delivered on inexpensive smartphones and tablets that students already own, and their use is less disruptive to a classroom environment than the donning of VR headsets. It is way too early to predict where AR educational tools are heading, but early efforts are promising.

With Facebook purchasing VR goggle maker Oculus Rift for $2 billion, Microsoft's expansive commitment in their HoloLens VR/AR headset, Google making many free virtual field trips to educational sites available online, and a large number of VR applications becoming available for smartphones and tablets, immersive education evangelist Aaron Walsh has called 2015 the beginning of "the age of immersion."[46] Time will tell whether his prophesy is ever realized, but the technology for immersive education tools will soon be available for educators to use at reasonable price points. If instructors can come up with compelling uses for the new capabilities these tools afford, immersive media may indeed join the pantheon of instructional media forms.

6

Making Sense of Media for Learning

Using new media forms for instruction has always been controversial. Even when the technology available to instructors was limited to reading and writing, there was debate as to whether new innovations should be used in teaching. Socrates thought his students should not be taught to write because he believed it would diminish their wisdom: "If men learn this, it will implant forgetfulness in their souls; they will cease to exercise memory because they rely on that which is written, calling things to remembrance no longer from within themselves, but by means of external marks. What you have discovered is a recipe not for memory, but for reminder."[1]

Similar cautions have followed the use of all kinds of technology and media for instruction. When inexpensive electronic calculators first appeared on the market, there was an uproar about using them in high school classrooms, with some educators fearing that their students, like Socrates's, would become too reliant on them and not learn how to do basic calculations in their head. In time, teachers began to realize that using calculators helped smooth the road toward learning higher-order mathematical skills, to the point that students must use one to complete the all-important SAT.[2] Not all efforts at introducing new technology and media into the classroom have been so warmly received as the calculator, largely because the calculator fits better into the usual flow of activity in the classroom. It may have displaced pencil, paper, and some mental energy, but its use fits well within customary classroom actions and roles and is relatively inexpensive. Not so with media and many other technology-based instructional tools.

Media in the classroom often changes the basic relationship between teacher and student. It is significantly more expensive to deploy and even more disruptive to the classroom's status quo. Instructors' primary role is no longer lecturing about a topic, and imparting content knowledge is relegated to the media—be it a film, video, or a filmstrip. Flipped-classroom methods, eLearning technologies such as massive open online courses, and augmented reality and simulation tools further stress the traditional teacher–student relationship and force instructors into novel situations with their students. This is not necessarily a bad thing, but it represents a challenge for successful and widespread adoption.

The Media *Is* the Method

In his influential 1964 book *Understanding Media*, Marshall McLuhan famously said that the "medium is the message."[3] His point has been often misconstrued, and that confusion was even parodied in Woody Allen's 1977 film *Annie Hall* (fig. 6.1) when Allen brings over McLuhan to admonish an academic blowhard behind him in a movie queue, saying, "You know nothing of my work!" But McLuhan believed that the nature of a media form has a profound influence on how that message is understood. Different forms of media had subtle but distinct attributes or affordances that

Figure 6.1. Marshall McLuhan's cameo in *Annie Hall* ("You know nothing of my work!").

combined with our cultural understandings to affect what was actually being communicated. McLuhan believed that the printed word encouraged more emphasis on visual sensory input, whereas communication from earlier oral cultures placed more weight on our sense of hearing, and therefore fundamentally changed the nature of the communication.

It is probably not too far-fetched for most people to believe that using different forms of media for instructional purposes is likewise affected by the media presenting content to the learner. For almost a century, educational researchers have examined the effects of using different forms of media on instruction through *media comparison studies*. In these studies, the same content is taught using two different forms of media (e.g., classroom instruction vs. television), and researchers compare the results of some sort of assessment. The method that yields the highest score is declared the winner, or when no significant difference is found, a tie.

The Clark-Kozma Debate

In 1983, University of Southern California educational researcher Richard Clark (fig. 6.2) ignited a decadelong debate with Robert Kozma from the University of Michigan (fig. 6.3) about the effectiveness of educational media that changed the direction of media research. Clark wrote, "The best current evidence is that media are mere vehicles that deliver instruction but do not influence student achievement any more than the truck that delivers our groceries causes changes in our nutrition."[4] He then called for a moratorium on further media comparison studies in favor of more nuanced investigations that explored the attributes of the media that enabled conditions to teach specific cognitive skills. Kozma disagreed with that position and the two men volleyed rebuttals and rebuttals to rebuttals in academic journals for a number of years, and they even went head to head at conference panels.

With five decades of inconclusive results from media studies, Clark's paper seemed to touch a nerve within the instructional technology community, which was grappling with the latest technology vying for attention: the personal computer. The response was rapid, and it became increasingly difficult to publish media comparison studies in academic journals or obtain grant funding for them. Clark was genuinely surprised at the response, saying "It is therefore a bit of a mystery why my restatement of the claim of 'no differences expected' a decade ago received so

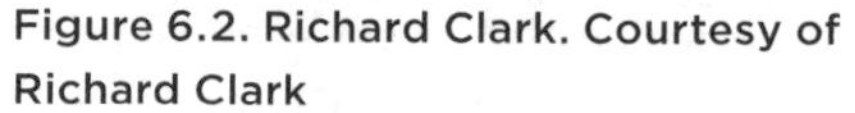

Figure 6.2. Richard Clark. Courtesy of Richard Clark

Figure 6.3. Robert Kozma. Courtesy of Robert Kozma

much attention. . . . I intended to stimulate discussion and I was not disappointed."[5]

The delivery truck metaphor seemed to resonate strongly with people, and gave them a tangible way to frame the issue of using technology for education. A student of Clark's, a physician, had come up with an earlier but far less effective rendition of the metaphor, suggesting that it would be ludicrous to expect different results from the same medicine delivered using a pill versus liquid form.[6] Clark had been thinking about this issue since the mid-1970s, and the provocative metaphor belied a more subtle plea for additional focused research on using media in education that went beyond simply comparing different types and seeing what worked better. He believed that it was impossible to untangle the effects of the teaching methodology from the media itself, which in his opinion was the less significant of the two factors, and urged researchers to focus on the instructional design of the overall interaction rather than simply effects of the media.[7]

He was not a Luddite, however, and admitted that media can have a potentially big impact on the cost of and access to instruction, just not on learning. Some form of media is necessary for instruction to occur, but Clark contended that the instructional method, not the media, was the secret sauce. Media had no real impact on learning itself, and any evidence to the contrary came from poorly designed research or was subject to the vested interests of the technology's purveyors.[8]

It took almost a decade, but in 1991, Robert Kozma wrote a rebuttal to Clark's article that became a minor media event in the world of educational technology, one that became required reading for educational technology students, even today. Both men were good friends and their discourse was always cordial. Kozma told me that the debate "made for good drama" and "was good fun," even though the difference in their positions was a matter of "splitting hairs."[9] Kozma felt that Clark's position, like Descartes's famous mind-body split, created an unnecessary schism between media and method. He believed the two factors were necessarily confounded and had an integral relationship, with both a part of the overall instructional design, saying, "In good designs, a medium's capabilities enable methods and the methods that are used take advantage of these capabilities. If media are going to influence learning, method must be confounded with medium. Media must be designed to give us powerful new methods, and our methods must take appropriate advantage of a medium's capabilities."[10] For example, video-based instruction has the ability to present information simultaneously through auditory and visual sensory channels. Investigating the effect that this dual coding has on learning would be useful in understanding how students learn using video.[11]

Clark and Kozma became the poster children for a larger schism that was occurring in the field of educational technology. Clark advocated for the dominant practitioners at that time, the followers of instructional design, and Kozma was more aligned with constructivist-leaning proponents of the learning sciences. Kozma took up the debate because he was one of the few from the latter contingent who actively talked with the instructional design people.[12]

Clark still stands by his position to this day. Some years ago, he offered a $5,000 prize to anyone who could show a true evidence-based learning benefit from using any specific type or combination of educational media that could not also be achieved with a different media mix. He had a few interested takers, but as of 2015, nobody has won the prize.[13] Kozma is characteristically more optimistic about media's potential positive effect on learning, especially as the digital convergence of disparate media types makes the issue more or less moot.[14]

Is "No Significant Difference" Such a Bad Thing?

When two methods of teaching are compared with one another in an experiment—for example, pitting traditional classroom instruction with instruction by television—a test or assessment often finds no significant difference in effectiveness, or NSD, and it is not unusual in the world of educational research. With such a finding the competing methodology is often dismissed, implying that it is not worth consideration or adoption. But a finding of NSD is more a reflection of the original research question rather than a blanket indictment against the competing method. In addition, assessments are blunt instruments in teasing out effectiveness, and they are notoriously bad at gauging less empirically oriented skills, such as those used for critical thinking.

The vast preponderance of studies on the use of media and technology do not show significant gains in student achievement, at least in regards to what the assessments are assessing. As far back as 1928, studies that measured the effectiveness of correspondence schools reported "no differences in test scores of college classroom and correspondence study students enrolled in the same subjects."[15] But there was a difference, perhaps not in *what* the students accomplished, but in *how* they accomplished their learning. The average correspondence school student was not able to attend or afford a traditional residential college education. He was male, 27 years old, worked a blue-collar job (often in a coal mine), and had a family to support. The student would study at home and "use the kitchen table as a desk, and often rock the cradle with one hand to keep baby quiet, while holding the lesson paper in the other."[16] The educational goals were finely focused, as one correspondence school executive explained, "the regular technical school or college aims to educate a man broadly; our aim, on the contrary, is to educate him only along some particular line."[17]

So perhaps we are looking for differences along the wrong dimension at this point. Obviously, it would be wonderful for researchers to report that their interventions have positive impacts on student learning, but there may be some other important factors to consider when judging a technology or media's worth. Even Richard Clark, who made a career of attacking these kinds of media comparison studies, agrees that "there are benefits to be gained from different media. The benefits are economic. If researchers and practitioners would switch their concerns to the economics of instruction, we would discover all manner of important cost contributions from the media."[18]

Another dimension to evaluate is the amount of time that technology and media-infused instruction takes relative to the traditional classroom. Many studies of intelligent tutoring systems[19] and programmed instruction tools[20] have shown a reduction in the time it takes to teach the same content by factors of two and three. I personally experienced the time savings in middle school when I took a paper-based programmed instruction tool to learn a semester's course in geometry in only a couple of months.

Education researcher Thomas Russell published a slim book in 1999 that looked at 355 media comparison studies, all of which yielded NSD results, and found a number of other potential benefits that technology and media could potentially offer beyond better economics, including that the number of course offerings could be substantially increased; courses could be provided rapidly to fulfill specific instructional needs; and, finally, that media and technology solutions could serve large and small populations with the same basic offering.[21]

The Role of Business in Education

I have no doubt that Edison truly believed in the potential of movies to change education when he uttered his prophetic but ultimately false prediction that movies would one day outnumber books in schools.[22] But the Wizard of Menlo Park also had a financial incentive to make that vision a reality, and I'm sure that the considerable investment the Edison Company made in creating and distributing educational films was not just for the benefit of the children. Business and education have long been partners in media, but that relationship can be complicated. It is important to openly recognize the self-interests that are necessarily involved in addition to the social obligations.

Educational experiences imply a trust relationship between the student and instructor, not unlike that between doctors and patients. Just as a physician prescribes treatment on the basis of her training, knowledge, and experience, the instructor guides learning down a path based on his own wisdom. Both the patient and the student need to trust that the doctor and instructor, respectively, have their best interests at heart. The introduction of money into the mix complicates this trust.

Thomas J. Foster, founder of the International Correspondence Schools (ICS), a wildly successful mail-based correspondence school during the early twentieth century, was keenly aware of that relationship. He had initially founded his business to provide safety instruction to eliminate accidents in the dangerous Pennsylvania coal mines, but whatever altruistic intentions he had, Foster unabashedly saw ICS as a business, one with a product it sold to customers. He wrote in 1906, "This is a commercial enterprise. It is necessarily so." But, he added, "The business is conducted for gain, but with gain as the motive influencing his teacher, the student fares as well as when he is the beneficiary of the state."[23]

The United States spends a tremendous amount of money on education. According to the World Bank, we spent over 5 percent of our gross domestic product[24] on education in 2011, or nearly $1 trillion, and each state spends an average of $10,700[25] per student in our public schools. There has been a constant drumbeat in the press, going back to the nineteenth century, that we are not getting our money's worth when we send our kids to public schools, creating a crisis of faith in the effectiveness of American education. The *Flexner Report*, published in 1910, revealed problems in early medical education, and it ushered in wholesale modernization in the way American doctors were trained.[26] In 1946, President Harry S. Truman's Commission on Higher Education concluded that American education levels "are still substantially below what is necessary," and the 1983 publication of the report *A Nation at Risk* criticized the state of American schools and spurred the use of technology in the classroom for the following two decades. [27]

But it was the Soviet Union's launch of the Sputnik satellite in 1957 that had the most reverberating effects on the American psyche, the military industrial complex, and the public educational system. The Soviet launch of a small satellite the size of a beach ball into a low earth orbit ignited fear that America was losing its technological dominance. The 184-pound Sputnik (Russian for "satellite," fig. 6.4) was visible to the naked eye as it orbited the earth every 96 minutes. "Words do not easily convey the American reaction to the Soviet satellite," Roger Launius, the former chief historian of NASA, wrote. "The only appropriate characterization that begins to capture the mood . . . involves the use of the word hysteria."[28]

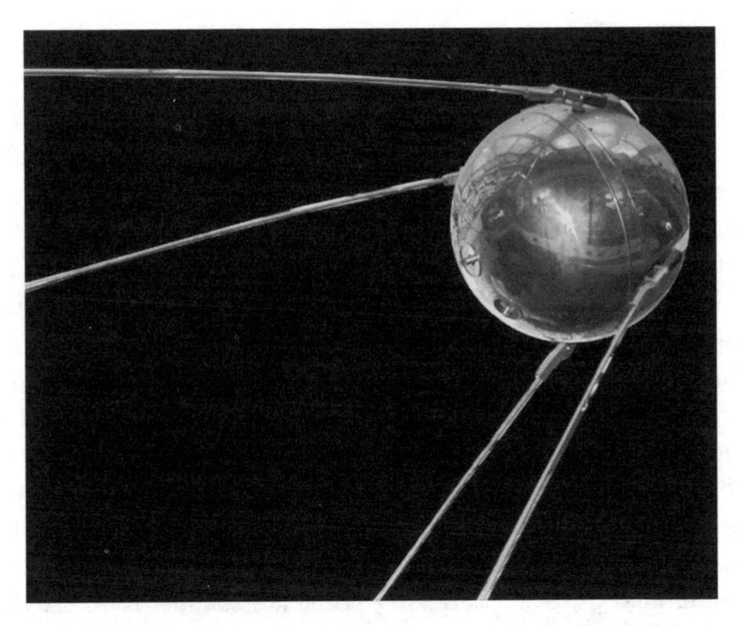

Figure 6.4. Model of the Soviet Sputnik I satellite. Courtesy of NASA

Fanning the flames of fear was the former assistant director of the Manhattan Project and "father" of the US Navy's atomic submarine fleet, Admiral Hyman George Rickover. Rickover viewed Sputnik's launch as a triumph of Russian education and saw the United States lagging behind in producing the scientists and engineers necessary to compete with them. "Education is the most important problem facing the United States today and only the massive upgrading of the scholastic standards of our schools will guarantee the future prosperity and freedom of the Republic." His call for a better educated populace to respond to that threat prompted a national outcry for a more effective and rigorous school system in America, and, not surprisingly, he advocated technology as a way toward this end.[29]

The Role of Venture Capital

Today, venture capital firms fund most educational technology and media companies. These firms aggregate large amounts of money from investors and strategically invest in a number of small start-up companies. They tend to invest with the assumption that only 10 percent of companies will succeed, but those successes typically yield a return on investment of at least 300 percent. Venture capital has built powerful and successful high-tech centers across the country, including those known as Silicon Alley (New York City), Silicon Beach (Los Angeles), Route 128 (Boston), and the most successful of them all, the San Francisco Bay Area's Silicon Valley. But there are also downfalls in this system. Companies that receive venture capital often give up a lot of control in exchange for funding. If a company does not deliver on its promises, that funding can dry up, leading to a change in the company's direction or even an end to the business altogether.

The education writer Kevin Carey wrote about a visit he made to a typical venture capital–funded education technology start-up in the heart of Silicon Valley. He described a chart on a whiteboard that resonated with me about how these companies choose their targets and why they just may be ultimately successful. The chart had four circles, each marked with the name of a sector and a dollar value representing the company's global sales potential: enterprise software, $0.3 trillion; e-commerce, $0.8 trillion; media and entertainment, $1.6 trillion; and the largest circle, education, worth $4.6 trillion. Carey's host explained that enterprise software and e-commerce were native-digital marketplaces, and media and entertainment were about half converted, but education stood alone as an almost entirely nondigital business. Venture capital firms have taken notice of the huge business opportunity and have been investing steadily in educational technology companies, increasing their stake from less than $200 million in 2008 to over $1.2 billion in 2013.[30]

Throughout the start-up community, companies are jockeying to get their own piece of that $4.6-trillion circle. To be sure, many will fail, but others will succeed and chip away at the overall market. Education's final surrender to the digital world won't be from one clear stroke of the sword of a single player, but more like death by a thousand cuts. This process will be slower in the K–12 world, because of their more entrenched bureaucracies, but colleges and universities will be vulnerable to this potentially disruptive force.

The media fury surrounding the rush of capital into the educational markets can have unintended consequences for institutions, which can be perceived as being slow to adopt the latest technologies. Teresa Sullivan, the president of my school, the University of Virginia, was summarily fired in a coup d'état in 2012 (and subsequently rehired because of faculty outrage), ostensibly because the board of visitors (trustees) perceived her not to be embracing online education rapidly enough.

Grant-Funded Projects

As someone who has raised large amounts of money over the years in federal and foundation grants to support educational research efforts, I should probably tread lightly here, but the grant-funded model for educational media projects has its flaws, mainly because grant funding ends at some point, and there have only been a few projects that have made the leap to a self-sustainable business model. The landscape is unfortunately strewn with excellent projects that in spite of good experimental results languish from a lack of funds to continue.

Commercial software distributors have offered successful university projects, such as the videodisc-based *Adventures of Jasper Woodbury* and the immersive *River City* project, but these products are often too expensive and not well marketed. Their academic parents usually do little or no continued development on them, and they typically die a slow death from a lack of attention. Like sharks, software and media-based projects need to keep moving to stay alive.

High Media Production Values Are a Tough Act to Follow

We live immersed in a media-rich environment, where the average cost to produce a 30-second television commercial was $354,000 in 2011,[31] with many ads costing millions of dollars to create and even more to air. Whatever opinions one might have about the ultimate value of commercial television programs, they are typically finely crafted productions created by professionals possessed with great communicative skills—and the money to accomplish them.

If we consider the three traditional genres that divide film and television content—*entertainment*, *educational*, and *instructional*—we have high expectations for those made for our entertainment. A single episode of a commercial television program may cost over a million dollars to produce, a typical PBS educational program costs in the hundreds of thousands, and an instructional program costs in the tens of thousands of dollars to create. These costs roughly correlate with the size of the audience each tier serves, and that audience size also provides the financial justification for the programming. Unfortunately, the perceived production quality also correlates with the amount spent to produce them.

Advancements in digital technology have theoretically made a flat playing field for all budget levels. Inexpensive cameras with high-definition quality and laptop-based editing and effects software can create programs that are qualitatively indistinguishable from expensive studio-produced content. Missing from this calculus, however, is the human factor. A virtual army of highly skilled writers, artisans, and professionals work in entertainment, and to a lesser extent in the educational sector, but the falloff is sharp when it comes to instructional media. Instructional programs typically make use of local talent, hastily written scripts, and overall lower production values, and they often fail to live up to the quality expectations among their audience.

This same dynamic occurs in the world of video gaming. Educators watch in awe how video games attract and hold the attention of players, and they have sought to create educational video games to capture some of that interest. But as with films and television, the expectations are set high by the entertainment segment of the gaming industry. The cost to produce a successful video game can rival a feature-film budget, costing many millions of dollars. The budgets of educational game developers are typically several orders of magnitude less, and so far they have not been able to create wildly successful instructional games.

The siren's lure of television's potential to scale education through mass broadcast is substantial. A 1972 production of Shakespeare's *Much Ado about Nothing* appeared on broadcast television to extremely poor ratings and much harsh criticism. Yet in spite of that public response and the subsequent chill it cast on future educational broadcasts for some time, more people experienced the play that night than had seen it performed in the 350 years since it was written.[32]

Can We Teach Using Media?

Groucho Marx once said, "I find television very educational. The minute someone turns it on, I go into the library and read a book."[33] Media and television, in particular, are mass communication tools and learning is an individual endeavor, so they are fundamentally at odds with one another. From a pedagogical perspective, using traditional media to teach violates much of what we have come to understand as good educational practices. It assumes a one-size-fits-all strategy that does not take into account the individual differences that exist among learners. Yet many products assume that all students come to instruction with the same amount of preexisting knowledge and learn at the same pace, offering one-way interaction with no real opportunity for feedback.

The experiment in American Samoa is a classic example of what *not* to do when introducing a technology like television into the classroom. Teachers were treated as implementers of a strategy delivered from on high, and they had no role in developing or customizing the curriculum to fit the changing daily needs of the classroom. The lack of flexibility to keep up with aired material, as "the TV kept coming," points out a major problem in the broadcast model for distribution of educational content in real-world classroom instruction.

Efforts for forcing media use in the classroom have not been effective because of the competing demands of central control versus teacher autonomy and pushback from schools and parents regarding the commercialization of education. They are odd bedfellows at best, snidely characterized by the media educator George Gordon as follows: "there has been no shortage of clowns, mountebanks,* jugglers and recreation directors who have thoroughly convinced themselves, that given the opportunity, they would turn out entirely competent to teach anything to anybody more than professional teachers. After all, they know how to sell snake oil on the midway, don't they?"[34]

Over the past 50 years, hundreds of academic studies have looked at the effectiveness of using instructional film, radio, and television in the classroom. Comparing students who used media against those who did not, the vast majority of experiments showed only modest gains, or the dreaded finding of no significant difference, typical in these types of comparison studies.[35] The confusing results from the research on *Sesame Street* are particularly ironic in light of the Children's Television Workshop's efforts to actively include research as a full component of their design and implementation plan. In the spirit of P. T. Barnum, however, it seems clear that some media can teach some people some of the time, the same way that some teachers can teach some students.

* A *mountebank* is a person who deceives others in order to trick them out of their money.

In the final analysis, it is probably less about how educational content is delivered, and more about how that instruction is designed and deployed. There is no question that employing media and technology can deliver instruction to many more people at a lower cost and ease of access than more traditional delivery vehicles. By looking at the previous ways people have tried in the past, we can learn from their successes and in our future efforts avoid where they've stumbled. As the legendary CBS journalist Edward R. Murrow said about broadcast television in 1958, "This instrument can teach, it can illuminate; yes, and it can even inspire. But it can only do so to the extent that humans are determined to use it to those ends. Otherwise it is merely wires and lights in a box."

Notes

Preface

1. J. Caldwell and A. Waring, "Some Chatham County Pottery Types and Their Sequence," *Southeastern Archaeological Conference Newsletter* 1, no. 5–6 (1939).

2. A. King, "From Sage on the Stage to Guide on the Side," *College Teaching* 41, no. 1 (1993): 30–35.

Chapter 1. Traditional Media

Epigraph: From the theme song from the 1960s television show *Mr. Ed*, written by Jay Livingston and Ray Evans and produced by Fairways Television. Accessed September 4, 2015, http://www.imdb.com/title/tt0054557.

1. A. Clark, *Lost Worlds of 2001* (New York: Roc, 1972), p. 189.

2. J. Kendall, *The Man Who Made Lists: Love, Death, Madness and the Creation of Roget's Thesaurus* (London: Putnam, 2008), pp. 3–17.

3. Ibid., p. 216.

4. P. Roget, "Explanation of an Optical Deception in the Appearance of the Spokes of a Wheel when Seen through Vertical Apertures," *Philosophical Transactions of the Royal Society of London* 115 (1825): 131–40.

5. J. Zacks, *Flicker: Your Brain on Movies* (New York: Oxford University Press, 2015), pp. 142–49.

6. Stanford University is named in memory of Stanford's only son, Leland Jr., who died of typhoid fever in 1891. Stanford founded the university, saying, "the children of California shall be our children." R. Cloud, *Education in California* (Stanford, CA: Stanford University Press, 1952), p. 114.

7. G. Clark, *Leland Stanford, War Governor of California: Railroad Builder and Founder of Stanford University* (Stanford, CA: Stanford University Press, 1931), p. 10.

8. Ibid., pp. 16–17.

9. N. Tutorow, *The Governor: The Life and Legacy of Leland Stanford, a California Colossus* (Spokane, WA: Clark, 2004), pp. 171–237.

10. Ibid., p. 437.

11. Ibid., pp. 401–55.

12. L. Mitchell, "The Man Who Stopped Time," *Stanford Alumni Magazine*, May/June 2001, https://alumni.stanford.edu/get/page/magazine/article/?article_id=39117.

13. E. Muybridge, *Animals in Motion* (New York: Dover, 1957), p. 13.

14. G. Hendricks, *Eadweard Muybridge: The Father of the Motion Picture* (New York: Grossman, 1975), p. 47.

15. E. Ball, *The Inventor and the Tycoon* (New York: Doubleday, 2013), p. 30.

16. Ibid., pp. 23–33.

17. M. Braun, *Picturing Time: The Work of Étienne-Jules Marey* (Chicago: University of Chicago Press, 1992), p. 2.

18. E. Marey, *La Machine animale: Locomotion terrestre et aérienne* (Paris: Baillie, 1873).

19. "The Paces of Horses," *Popular Science* 6 (1874): 130.

20. Muybridge, *Animals in Motion*, p. 49.

21. Hendricks, *Eadweard Muybridge*, p. 102.

22. Ball, *The Inventor and the Tycoon*, pp. 311–23.

23. Ibid., pp. 15–19.

24. N. Baldwin, *Edison: Inventing the Century* (New York: Hyperion, 1995), p. 211.

25. "Living with a Genius," *Reader's Digest*, March (1930): 1042–44.

26. A. Williams, *Republic of Images: A History of French Filmmaking* (Cambridge, MA: Harvard University Press, 1992), pp. 20–22.

27. G. Hendricks, *The Edison Motion Picture Myth* (Los Angeles: University of California Press, 1961), p. 14.

28. R. Phillips, *Edison's Kinetoscope and Its Films: A History to 1896* (Westport, CT: Greenwood, 1997), pp. 21–27.

29. Hendricks, *Edison Motion Picture Myth*, pp. 108–11.

30. F. Dyer and T. Martin, *Edison: His Life and Inventions* (New York: Harper, 1920), pp. 543–48.

31. Phillips, *Edison's Kinetoscope*, pp. 60–63.

32. Ibid., p. 47.

33. R. Allen, "Moving Picture Exhibition in Manhattan 1906–1912: Beyond the Nickelodeon," *Cinema Journal* 18, no. 2 (1979): 3–4.

34. "The Electric Tachyscope," *Scientific American* 16 (1889): 301.

35. W. Carlson and G. Gorman, "Understanding Invention as a Cognitive Process: The Case of Thomas Edison and Early Motion Pictures," *Social Studies of Science* 20 (1990): 406–8.

36. Hendricks, *The Edison Motion Picture Myth*, p. 92.

37. C. Musser, "Nationalism and the Beginnings of Cinema: The *Lumiere cinematographe* in the US, 1896–1897," *Historical Journal of Film, Radio* 19, no. 2 (1999): 149–76.

38. F. Smith, "The Evolution of the Motion Picture: VI. Looking into the Future with Thomas A. Edison," *New York Dramatic Mirror*, July 9, 1913, p. 24.

39. W. Dickson and A. Dickson, *History of the Kinetograph, Kinetoscope, and Kinetophongraph* (New York: Albert Bunn, 1895), p. 51.

40. P. Saettler, *The Evolution of American Educational Technology* (Greenwich, CT: Information Age, 2004), p. 96.

41. G. Kleine, *Catalogue of Educational Motion Pictures* (New York: George Kleine, 1910), p. 1.

42. Saettler, *Evolution of American Educational Technology*, pp. 98–99.

43. "National Archives and Records Service Film-Vault Fire at Suitland, MD," *Internet Archive*, accessed February 19, 2016, http://archive.org/stream/nationalarchivesoounitrich/nationalarchivesoounitrich_djvu.txt.

44. F. Freeman, "Requirements of Education with Reference to Motion Pictures," *School Review* 31, no. 5 (1923): 340–42.

45. F. McClusky, "The Nature of the Educational Film," *Hollywood Quarterly* 2, no. 4 (1947): 372.

46. A. Geoff, *Academic Films for the Classroom: A History* (Jefferson, NC: McFarland, 2010), pp. 54–56.

47. Ibid., pp. 15–22.

48. D. Orgernon, M. Orgernon, and D. Streible, *Learning with the Lights Off: Educational Film in the United States* (Oxford: Oxford University Press, 2012), p. 35.

49. E. Barnouw, *A History of Broadcasting in the United States* (New York: Oxford University Press, 1968), p. 30.

50. O. Dunlap, *Marconi: The Man and His Wireless* (New York: Macmillan, 1937), pp. 11–16.

51. E. Barnouw, *Tube of Plenty: The Evolution of American Television* (Oxford: Oxford University Press, 1992), p. 9.

52. P. Heyer, *Titanic Century: Media, Myth, and the Making of a Cultural Icon* (Santa Barbara, CA: ABC-CLIO, 2012), p. 51.

53. D. Marconi, *My Father, Marconi* (Toronto, ON: Guernica Editions, 2001), p. 165.

54. B. Fabos, *Wrong Turn on the Information Superhighway: Education and the Commercialization of the Internet* (New York: Teachers College Press, 2004), pp. 8–11.

55. W. Bianchi, "Education by Radio: America's Schools of the Air," *TechTrends: Linking Research and Practice to Improve Learning* 52, no. 2 (2008): 36–44.

56. Ibid., p. 39.

57. W. Schramm, *Big Media, Little Media: Tools and Technologies for Instruction* (Beverly Hills, CA: Sage, 1977), pp. 33–58.

58. E. Schwartz, *The Last Lone Inventor* (New York: HarperCollins, 2002), pp. 13–20.

59. G. Edgerton, *The Columbia History of American Television* (New York: Columbia University Press, 2007), pp. 39–41.

60. Barnouw, *Tube of Plenty*, pp. 77–78

61. D. Stashower, *The Boy Genius and the Mogul: The Untold Story of Television* (New York: Broadway Books, 2002), p. 143.

62. Edgerton, *Columbia History of American Television*, pp. 43–46.

63. A. Abramson, *Zworykin, Pioneer of Television* (Chicago: University of Illinois Press, 1995), p. 78.

64. Edgerton, *Columbia History of American Television*, p. 48.

65. E. Schwartz, *The Last Lone Inventor* (New York: HarperCollins, 2002), pp. 70–72.

66. Stashower, *The Boy Genius and the Mogul*, p. 244.

67. Schwartz, *Last Lone Inventor*, p. 265.

68. Saettler, *Evolution of American Educational Technology*, pp. 359–65.

69. J. Murphy and R. Gross, *Learning by Television* (New York: Fund for the Advancement of Education, 1966), pp. 10–12.

70. *Teaching by Television: A Report from the Ford Foundation and the Fund for the Advancement of Education*, Technical Report (New York: Fund for the Advancement of Education and Ford Foundation, 1961), p. 29.

71. Ibid., p. 15.

72. J. Bergmann and A. Sams, *Flip Your Classroom: Reach Every Student in Every Class Every Day* (Eugene, OR: International Society for Technology in Education, 2012).

73. W. Schramm, *Bold Experiment: The Story of Educational Television in American Samoa* (Stanford, CA: Stanford University Press, 1981), pp. 6–11.

74. Ibid., pp. 16–25.

75. H. Lee, "Planning Communications Facilities for Public Education," *Annals of the New York Academy of Sciences* 142 (1967): 531–32.

76. Schramm, *Bold Experiment*, pp. 35–55.

77. Ibid., p. 77.

78. Ibid., pp. 74–80.

79. M. Davis, *Street Gang: The Complete History of Sesame Street* (New York: Viking, 2008), p. 11.

80. Ibid., pp. 12–16.

81. Ibid., p. 26.

82. Ibid., p. 27.

83. C. Odell, *Women Pioneers in Television: Biographies of Fifteen Industry Leaders* (Jefferson, NC: McFarland, 1997), pp. 68–71.

84. R. Morrow, *Sesame Street and the Reform of Children's Television* (Baltimore: Johns Hopkins University Press, 2006), pp. 48–62.

85. S. Fisch, *Children's Learning from Educational Television* (Mahwah, NJ: Laurence Erlbaum, 2004), pp. 17–19.

86. "Who Wants to Live on Sesame Street?" *Science News* 103, no. 12 (1973): 183.

87. T. Cook, H. Appleton, R. Conner, A. Shaffer, G. Tamkin, and S. Weber, *Sesame Street Revisited* (New York: Russell Sage Foundation, 1975), pp. 2–21.

88. M. Kearney and P. Levine, "Early Childhood Education by MOOC: Lessons from Sesame Street," Working Paper 21229, National Bureau of Economic Research, Cambridge, MA, 2015, http://www.nber.org/papers/w21229.

89. Fabos, *Wrong Turn on the Information Superhighway*, pp. 16–19.

Chapter 2. Interactive Media

Epigraph: From a 1931 song written by Duke Ellington and Irving Mills.

1. D. Jonassen, "Interactive Lesson Designs: A Taxonomy," in *Interactive Video*, vol. 1, *The Educational Technology Anthology Series* (Englewood Cliffs, NJ: Educational Technology, 1989), pp. 19–29.

2. J. Lipson, "Design and Development of Programs for the Videodisc," *Journal of Educational Technology Systems* 9, no. 3 (1980): 278.

3. R. Mayer, "Cognitive Theory of Multimedia Learning," in *Cambridge Handbook of Multimedia Learning*, ed. R. Mayer (New York: Cambridge University Press, 2005), pp. 31–41.

4. G. Miller, "The Magical Number Seven, Plus or Minus Two: Some Limits on Our Capacity for Processing Information," *Psychological Review* 63, no. 2 (1956): 81–97.

5. J. Bransford, A. Brown, and R. Cocking, *How People Learn: Brain, Mind, Experience, and School* (Washington, DC: National Academy Press, 2000).

6. Ibid., pp. 51–78.

7. D. McLean, *Restoring Baird's Image* (London: Institute of Electrical Engineers, 2000), p. xvi.

8. Ibid., pp. 26–53.

9. L. Forsdale, *8mm in Education: Status and Products* (Washington, DC: Office of Education, Bureau of Research, 1968), pp. 2–13, http://eric.ed.gov/?id=ED039707.

10. R. Butler, "The Development and Demise of 8mm Loops in America," in *Vision Quest: Journeys toward Visual Literacy. Selected Readings from the Annual Conference of the International Visual Literacy Association* (Cheyenne, WY, 1997), pp. 191–96, http://eric.ed.gov/?id=ED408967.

11. N. Stafford, *"Dial-Access" as an Instructional Medium* (Washington, DC: Office of Education, Bureau of Research, 1970), pp. 4–10, http://eric.ed.gov/?id=ED039712.

12. "Business Outlook: Interactive Videodiscs Turn the Corner in 1983," *High Technology*, November 1983, p. 60.

13. E. Sigel, M. Schubin, and P. Merrill, *Videodiscs: The Technology, the Application and the Future* (White Plains, NY: Knowledge Industry, 1980), pp. 15–37.

14. P. Bosselmann, "The Berkeley Environmental Simulation Laboratory: A 12 Year Anniversary," *Berkeley Planning Journal* 1, no. 1 (1984): 150–52.

15. D. Bennahum, "Mr. Big Idea," *New York Magazine*, November 13, 1995, p. 76.

16. G. Hamel, *The Future of Management* (Boston: Harvard Business School Press, 2007), p. 113.

17. S. Brand, *The Media Lab: Inventing the Future at MIT* (New York: Penguin, 1988), p. 6.

18. N. Negroponte, *Being Digital* (New York: Knopf, 1995), pp. 65–66.

19. R. Mohl, personal communication, January 24, 2015.

20. A. Lippman, "Movie-Maps: An Application of the Optical Videodisc to Computer Graphics," *ACM SIGGRAPH Computer Graphics* 14, no. 3 (1980): 32–42.

21. R. Mohl, "Massachusetts Institute of Technology: Surrogate Travel," in *Handbook of Interactive Video*, ed. S. Floyd and B. Floyd (White Plains, NY: Knowledge Industry, 1980), pp. 150–55.

22. Brand, *Media Lab*, p. 141.

23. T. Nelson, *Computer Lib: Dream Machines.* (Redmond, WA: Tempus Books, 1987), p. 66.

24. D. Heuston, personal communication, January 26, 2015.

25. D. Heuston and J. Parkinson, *The Third Source: A Message for Hope in Education* (Salt Lake City, UT: Waterford Institute, 2011), pp. 95–96.

26. C. Victor Bunderson, personal communication, February 3, 2015.

27. D. Heuston, personal communication, January 26, 2015.

28. E. Schneider and J. Bennion, *Videodiscs* (Englewood Cliffs, NJ: Educational Technology, 1981), 33–34.

29. R. Mendenhall, personal communication, January 18, 2015.

30. C. Bunderson, "Instructional Strategies for Videodisc Courseware: The McGraw Hill Disc," *Journal of Educational Technology Systems* 8, no. 3 (1979): 207–10.

31. C. Bunderson, B. Biallio, and J. Olsen, "Instructional Effectiveness of an Intelligent Videodisc in Biology," *Machine-Mediated Learning* 1, no. 2 (1980): 175–215.

32. Sigel et al., *Videodiscs*, pp. 15–37.

33. T. Hasselbring, personal communication, July 9, 2015.

34. K. Reusser, "Problem Solving beyond the Logic of Things: Contextual Effects on Understanding and Solving Word Problems," *Instructional Science* 17 (1988): 324.

35. S. Spielberg (director), *Raiders of the Lost Ark*, Paramount Pictures, 1981.

36. T. Hasselbring, personal communication, July 9, 2015.

37. J. Bransford, R. Sherwood, T. Hasselbring, C. Kinzer, and S. Williams, "Anchored Instruction: Why We Need It and How Technology Can Help," in *Cognition, Education and Multimedia: Exploring Ideas in High Technology*, ed. D. Nix and R. Spiro (Hillsdale, NJ: Laurence Erlbaum, 1990), pp. 129–31.

38. Cognition and Technology Group at Vanderbilt, *The Jasper Project: Lessons in Curriculum, Instruction, Assessment and Professional Development* (Mahwah, NJ: Laurence Erlbaum, 1997), p. vii.

39. Ibid., p. 4.

40. Cognition and Technology Group at Vanderbilt, "The Jasper Project: An Exploration of Issues in Learning and Instructional Design," *Educational Technology Research and Development* 40, no. 1 (1992): 65–80.

41. Cognition and Technology Group at Vanderbilt, *Jasper Project*, p. 45.

42. Jasper Project, "The Jasper Experiment: Using Video to Furnish Real-World Problem-Solving Contexts," *Arithmetic Teacher* 40, no. 8 (1993): 474–78.

43. D. Viadero, "The Adventures of Jasper Woodbury," *Education Week*, February 1, 1992.

44. B. Barron and R. Kantor, "Tools to Enhance Math Education: The Jasper Series," *Communications of the ACM* 36, no. 5 (1993): 52.

45. T. Hasselbring, personal communication, July 9, 2015.

46. Cognition and Technology Group at Vanderbilt, *Jasper Project*, pp. 158–62.

47. Barron and Kantor, "Tools to Enhance Math Education," pp. 52–54.

48. Rossetti Archive, accessed February 20, 2016, www.rossettiarchive.org.

49. Cognition and Technology Group at Vanderbilt, *Jasper Project*, pp. 62–78.

50. Ibid., pp. 136–42.

51. Cognition and Technology Group at Vanderbilt, "The Adventures of Jasper Woodberry," accessed February 20, 2016, http://jasper.vueinnovations.com.

52. T. Hasselbring, personal communication, July 9, 2015.

53. K. Wood and R. Woolley, *An Overview of Videodisc Technology and Some Potential Applications in the Library, Information, and Instructional-Sciences* (Washington, DC: National Institute of Education, 1980), p. 21, http://eric.ed.gov/?id=ED206328.

54. P. Gough, "ABC/NEA Schooldiscs: Will Apparent Problems Cancel Out Potential?," *Phi Delta Kappan* 62, no. 4 (1980): 247–49.

55. S. Corwin and R. Perlin, "A Videodisc Resource for Interdisciplinary Learning: American Art from the National Gallery of Art," *Art Education* 48, no. 3 (1995): 17–24.

56. R. Perlin, personal communication, June 15, 2015.

57. Ibid.

58. J. Bosco, "Interactive Video: Educational Tool or Toy?," in *Interactive Video*, ed. J. Bosco (Englewood Cliffs, NJ: Educational Technology, 1989), pp. 3–9.

59. K. Kritch, D. Bostow, and R. Dedrick, "Level of Interactivity of Videodisc Instruction on College Students' Recall of AIDS Information," *Journal of Applied Behavioral Analysis* 1, no. 28 (1995): 85–86.

Chapter 3. Hypermedia

Epigraph: S. Coleridge, "Kublai Khan," in *Kublai Khan*, ed. John Man (New York: Random House (1816 / 2012), p. 390.

1. D. Allen, "A Conversation with Nicholas Negroponte," *Videography* 6, no. 10 (1981): 46.
2. Z. Pascal, *Endless Frontier: Vannevar Bush, Engineer of the Twentieth Century* (New York: Free Press, 1997), pp. 20–84.
3. V. Bush, Profile-Tracer. US Patent 1,048,649, filed February 2, 1912, and issued December 31, 1912.
4. Pascal, *Endless Frontier*, p. 274.
5. Ibid., pp. 28–88.
6. Ibid., pp. 88–201.
7. V. Bush, "As We May Think," *Atlantic Monthly*, July 1945, p. 101.
8. Ibid., pp. 101–9.
9. Ibid., p. 101.
10. Pascal, *Endless Frontier*, pp. 268–69.
11. Ibid., pp. 101–9.
12. Ibid., p. 108.
13. D. Engelbart to Vannevar Bush, May 24, 1962, accessed February 20, 2016,http://web.stanford.edu/dept/SUL/library/extra4/sloan/mousesite/EngelbartPapers/LetterToVBush.html.
14. H. Rheingold, *Tools for Thought: The History and Future of Mind-Expanding Technology* (Cambridge: MIT Press, 2000), pp. 174–203.
15. D. Engelbart, *Augmenting Human Intellect: A Conceptual Framework*, AFOSR-3223 (Washington, DC: Air Force Office of Scientific Research, 1963), p. 49, www.dougengelbart.org/pubs/papers/scanned/Doug_Engelbart-AugmentingHumanIntellect.pdf.
16. Ibid., pp. 1–131.
17. Rheingold, *Tools for Thought*, pp. 174–203.
18. "The Mother of All Demos, Presented by Douglas Engelbart (1968)," YouTube video, 1:40:52, from the Fall Joint Computing Conference in San Francisco, December 9, 1968, posted by "MarcelVEVO," July 9, 2012, https://www.youtube.com/watch?v=yJDv-zdhzMY.
19. T. Bardini, *Bootstrapping: Douglas Engelbart, Coevolution, and the Origins of Personal Computing* (Stanford, CA: Stanford University Press, 2000), pp. 168–72.
20. M. Gladwell, "Creation Myth: Xerox PARC, Apple, and the Truth about Innovation," *New Yorker*, May 16, 2011.
21. M. Wittier (director) and D. Adams (writer), *Hyperland*, British Broadcasting Corporation, September 21, 1990.
22. T. Nelson, *Possiplex: Movies, Intellect, Creative Control, My Computer Life and the Fight for Civilization* (Hackettstown, NJ: Mindful Press, 2010), p. 35.
23. B. Barnet, *Memory Machines: The Evolution of Hypertext* (London: Anthem Press, 2013), p. 66.
24. S. Rushdie, *Haroun and the Sea of Stories* (New York: Penguin, 1991), p. 71.
25. Rheingold, *Tools for Thought*, p. 299.

26. Nelson, *Possiplex*, p. 24.

27. G. Wolf, "The Curse of Xanadu," *Wired Magazine*, June 1, 1995, http://www.wired.com/1995/06/xanadu/.

28. Nelson, *Possiplex*, p. 9.

29. T. Nelson, "Complex Information Processing: A File Structure for the Complex, the Changing and the Indeterminate, in *Proceedings of the 1965 20th National Conference*, ed. Lewis Winner (New York: ACM, 1965), p. 96.

30. "Last Word: The Babbage of the Web," *Economist*, December 7, 2000, http://www.economist.com/node/442985.

31. Nelson, *Possiplex*, p. 180.

32. Barnet, *Memory Machines*, pp. 31–33.

33. Wolf, "The Curse of Xanadu."

34. Barnet, *Memory Machines*, p. 88.

35. Theodor Holm Nelson, "Curriculum Vitae," accessed July 10, 2015, http://hyperland.com/TNvita.

36. P. Lemmons, "The Interview: The Macintosh Design Team," *Byte Magazine* 9, no. 2 (1984): 63.

37. W. Isaacson, *Steve Jobs* (New York: Simon & Schuster, 2011), p. 94.

38. Ibid., p. 154.

39. K. Woolsey, *Learning to Run: A Firsthand Account of the Early Days of Interactive Multimedia at Apple* (forthcoming), pp. 46–71.

40. "HyperCard," Internet Archive video, 26:50, introduction to HyperCard, from the television show *Computer Chronicles*, Steward Cheifet Productions, 1987, posted July 14, 2004, https://archive.org/details/CC501_hypercard.

41. Barnet, *Memory Machines*, p. 129.

42. K. Frenkel, "The Next Generation of Interactive Technologies," *Communications of the ACM* 32, no. 7 (1989): 880.

43. M. Mills and R. Pea, "Mind and Media in Dialog: Issues in Multimedia Composition," in *Full-Spectrum Learning Conference Report*, ed. K. Hooper and S. Ambron (Cupertino, CA: Apple Computer, 1989), pp. 14–37.

44. Woolsey, *Learning to Run*, p. 295.

45. Ibid., p. 275.

46. L. Kahney, "HyperCard: What Could Have Been?" *Wired Magazine*, August 14, 2002, http://archive.wired.com/gadgets/mac/commentary/cultofmac/2002/08/54370.

47. D. Willingham, "Learning Styles FAQ," accessed July 20, 2015, http://www.danielwillingham.com/learning-styles-faq.html.

48. R. Spiro, R. Coulson, P. Feltovich, and D. Anderson, *Cognitive Flexibility Theory: Advanced Knowledge Acquisition in Ill-Structured Domains*, Technical Report No. 441, ERIC No. ED302821 (Cambridge, MA: Bolt, Barenak, and Newman, 1988), p. 6.

49. R. Mayer, *Multimedia Learning* (New York: Cambridge University Press, 2009), pp. 1–25.

50. R. Mayer, "Introduction to Multimedia Learning," in *Cambridge Handbook of Multimedia Learning*, ed. R. Mayer (New York: Cambridge University Press, 2005), pp. 6–7.

51. D. Norman, *The Design of Everyday Things* (New York: Doubleday, 1990).

52. K. Hooper Woolsey, personal communication, February 9, 2015.

53. J. Maloney, "Apple's Multimedia Lab: A Linear History," accessed July 15, 2015, http://www.caruso.com/work/dm-index/digital-media-september-1992/apples-multimedia-lab-a-linear-history.

54. Woolsey, *Learning to Run*, p. 20.

55. Ibid., pp. 21–33.

56. M. Bradsher, "The Teacher as Navigator," in *Learning with Interactive Media*, ed. S. Ambron and K. Hooper (Redmond, WA: Microsoft Press, 1990).

57. H. McLellan, *Situated Learning Perspectives* (Englewood Cliffs, NJ: Educational Technology, 1996), pp. 5–17.

58. R. Campbell and P. Hanlon, "Grapevine," in *Learning with Interactive Media* (Redmond, WA: Microsoft Press, 1990).

59. Maloney, *Apple's Multimedia Lab*.

60. N. Negroponte, *Being Digital* (New York: Knopf, 1995), p. 18.

61. A. Noll, "Scanned-Display Computer Graphics," *Communications of the ACM* 14, no. 3 (1971): 143–50.

62. The Moving Picture Experts Group Web site, accessed July 17, 2015, http://mpeg.chiariglione.org.

63. W. Gates, "Foreword," in *CDROM: The New Papyrus* (Redmond, WA: Microsoft Press, 1986), p. xiii.

64. A. Barlow, *The DVD Revolution* (Westport, CT: Praeger, 2005), pp. 17–18.

65. C. Netiva and M. Paprzycki, "Educational Software: Are We Approaching It the Right Way?," *Technology and Teacher Education Annual* (1996): 869–73.

66. C. Shuler, *Where in the World Is Carmen Sandiego? The Edutainment Era: Debunking Myths and Sharing Lessons Learned* (New York: Joan Ganz Cooney Center at Sesame Workshop, 2012).

67. E. Klopfer and S. Osterweil, "The Boom and Bust and Boom of Educational Games," in *Transactions on Edutainment IX*, vol. 7544, *Lecture Notes in Computer Science*, ed. Z. Pan et al. (New York: Springer, 2013), p. 291.

68. B. Stein, "Becoming Book-Like: Bob Stein and the Future of the Book," *Kairos* 15, no. 2 (2011): 5, http://kairos.technorhetoric.net/15.2/interviews.

69. D. Visel, "Mao, King Kong, and the Future of the Book: Bob Stein in Conversation with Dan Visel," *Triple Canopy*, July 23, 2010, https://canopycanopycanopy.com/issues/9/contents/mao__king_kong__and_the_future_of_the_book.

70. Ibid., p. 5.

71. S. Greenstein and M. Devereux, *The Crisis at Encyclopaedia Britannica* (Evanston, IL: Kellogg School of Management, Northwestern University, 2009), p. 7.

72. Ibid., pp. 10–14.

73. Ibid., p. 6.

74. L. Mitgang, *Big Bird and Beyond: The New Media and the Markle Foundation* (New York: Fordham University Press, 2000), pp. 184–85.

75. K. Hafner and A. Rogers, "How Now Voyager?" *Newsweek* October, 9, 1995, p. 67.

76. B. Vershbow, "Beethoven's Ninth Symphony CD Companion," October 20, 2005, http://futureofthebook.org/next/text/precursors.

77. Visel, "Mao, King Kong, and the Future of the Book", p. 12.

78. Institute for the Future of the Book, "Mission Statement," accessed July 10, 2015, http://futureofthebook.org/mission.html.

Chapter 4. Cloud Media

Epigraph: D. Marquis, *Archy and Mehitabel* (New York: Knopf Doubleday, 2012), p. 54.

1. T. Kidd, *Online Education and Adult Learning: New Frontiers for Teaching Practices* (Hershey, PA: Information Science Reference, 2010), pp. 35–36.
2. D. Comer, *Computer Networks and Internets* (New York: Prentice Hall, 2009), p. 99.
3. L. Cremin, *The Transformation of the American School* (New York: Knopf, 1961), pp. 110–13.
4. L. Benjamin, personal communication, August 25, 2013.
5. C. Ferster and S. Culbertson, *Behavior Principles* (Englewood Cliffs, NJ: Prentice-Hall, 1982), pp. 28–35.
6. B. F. Skinner, "Reflections on a Decade of Teaching Machines," in *Teaching Machines and Programmed Learning*, ed. R. Glaser (Washington, DC: National Education Association, 1965), p. 7.
7. Plato, *Meno*, translated by Benjamin Jowett, Internet Classics Archive, accessed February 13, 2016, http://classics.mit.edu/Plato/meno.html.
8. J. Holland, "Teaching Machines: An Application of Principles from the Laboratory," *Journal of the Experimental Analysis Behavior* 3, no. 4 (1960): 279.
9. E. Bozhovich, "Zone of Proximal Development: The Diagnostic Capabilities and Limitations of Indirect Collaboration," *Journal of Russian and East European Psychology* 47, no. 6 (2009): 49.
10. P. Young, *Teaching, Learning and the Mind* (Boston; Houghton-Mifflin, 1973), pp. 94–95.
11. J. Bick, "The Microsoft Millionaires Come of Age," *New York Times*, May 29, 2005, http://www.nytimes.com/2005/05/29/business/yourmoney/the-microsoft-millionaires-come-of-age.html.
12. 4Video, "The History and Future of Real Networks," *Internet Video Magazine*, accessed July 30, 2015, https://web.archive.org/web/20040228165516/http://internetvideomag.com/Articles-2004/022324Real.htm.
13. US History in Context, "RealNetworks, Inc.," in *International Directory of Company Histories*, vol. 53, ed. Tina Grant (Detroit: St. James Press, 2003), pp. 280–82.
14. E. Dinger, "YouTube, Inc.," in *International Directory of Company Histories*, vol. 90, ed. Tina Grant (Detroit: St. James Press, 2008), pp. 443–46.
15. J. Hopkins, "Surprise! There's a Third YouTube Co-Founder," *USA Today*, October 11, 2006, http://usatoday30.usatoday.com/tech/news/2006-10-11-youtube-karim_x.htm.
16. "Jawed Karim, Illinois Commencement 2007, pt2," YouTube video, 7:18, commencement address at the University of Illinois, part 2, posted by "Computer Science at Illinois," June 5, 2007, www.youtube.com/watch?t=84&v=24yglUYbKXE.
17. D. Altraide, "The Surprising History of YouTube!," YouTube video, 11:45, examination of the size and founding of YouTube, posted by "ColdFusion," February 4, 2105, https://www.youtube.com/watch?v=P4dT-lW926o.

18. "Statistics," YouTube Web site, accessed August 22, 2015, https://www.youtube.com/yt/press/statistics.html.

19. C. Ghilani and T. Seybert, "Enhancing Courses with Video Lessons," *Surveying and Land Information Science* 74, no. 1 (2015): 28–33.

20. J. Udell, "Name That Genre," *Jon Udell's Weblog* (blog), *InfoWorld*, November 15, 2005, https://web.archive.org/web/20050306092814/http://weblog.infoworld.com/udell/2004/11/15.html#a1114.

21. TechSmith Web site, accessed February 14, 2015, https://www.techsmith.com.

22. M. McLuhan, *Understanding Media: The Extensions of Man* (New York: McGraw-Hill, 1964), p. 289.

23. C. Goldberg, "Auditing Classes at MIT on the Web and Free," *New York Times*, April 4, 2001.

24. "Site Statistics," MIT OpenCourseWare Web site, accessed February 14, 2015, http://ocw.mit.edu/about/site-statistics.

25. The Conversations Network Web site, accessed July 29, 2013, http://web.archive.org/web/20130729200341id_/http://www.conversationsnetwork.org.

26. TED Web site, accessed August 30, 2015, http://ed.ted.com.

27. F. Rosell-Aguilar, "Podcasting for Language Learning through iTunes U: The Learner's View," *Language Learning and Technology* 17, no. 3 (2013): 74–93.

28. Apple Education Web site, accessed August 30, 2015, http://www.apple.com/education/ipad/itunes-u.

29. G. Bull, B. Ferster, and W. Kjellstrom, "Connected Classroom—Inventing the Flipped Classroom," *Learning and Leading with Technology* 40, no. 1 (2012): 10.

30. *The Four Pillars of F-L-I-P* (South Bend, IN: Flipped Learning Network, 2014), p. 9, http://flippedlearning.org/cms/lib07/VA01923112/Centricity/Domain/46/FLIP_handout_FNL_Web.pdf.

31. D. Berrett, "How 'Flipping' the Classroom Can Improve the Traditional Lecture," *Chronicle of Higher Education*, February 19, 2012, pp. 1–14.

32. *Four Pillars of F-L-I-P*, p. 8.

33. J. Bishop and M. Verleger, "The Flipped Classroom: A Survey of the Research," paper presented at the 120th American Society of Engineering Education Annual Conference and Exposition, Atlanta, GA, June 23–26, 2013.

34. M. Noer, "Reeducating Education," *Forbes*, November 19, 2012, p. 98.

35. "Press Room," Khan Academy Web site, accessed August 20, 2015, https://khanacademy.zendesk.com/hc/en-us/articles/202483630-Press-room.

36. S. Khan, *The One World Schoolhouse: Education Reimagined* (New York: Twelve, 2012), pp. 16–25.

37. C. Thompson, "How the Khan Academy Is Changing the Rules of Education," *Wired Magazine*, July 15, 2011, http://www.wired.com/2011/07/ff_khan.

38. B. Bloom, "The 2-sigma Problem: The Search for Methods of Group Instruction as Effective as One-to-One Tutoring," *Educational Researcher* 13, no. 6 (1984): 4–16.

39. B. Means, Y. Toyama, R. Murphy, M. Bakia, and K. Jones, *Evaluation of Evidence-Based Practices in Online Learning: A Meta-Analysis and Review of Online Learning Studies* (Washington, DC: US Department of Education, 2009), http://files.eric.ed.gov/fulltext/ED505824.pdf.

40. College Board, "Free, Personalized Practice for the Redesigned SAT® Now Available for All Students on KhanAcademy.org," press release, accessed August 23, 2015, https://www.collegeboard.org/releases/2015/personalized-practice-for-the-redesigned-sat-now-available-on-khanacademy-org.

41. "About the Standards," Common Core State Standards Initiative Web Site, accessed August 23, 2015, http://www.corestandards.org/about-the-standards.

42. W. Harms and I. DePencier, *100 Years of Learning at the University of Chicago Laboratory School* (Chicago: University of Chicago Laboratory Schools, 1996), http://www.ucls.uchicago.edu/about-lab/current-publications/history/index.aspx.

43. "Tutoring and Test Preparation Franchises in the US: Market Research Report," IBISWorld Web site, accessed August 12, 2015, www.ibisworld.com/industry/tutoring-test-preparation-franchises.html.

44. College Board, "Free, Personalized Practice."

45. L. Gordon, "College Board, Khan Academy Team Up to Offer Free SAT Prep Program," *Los Angeles Times*, June 2, 2015, www.latimes.com/local/education/la-me-sat-prep-20150602-story.html.

46. L. Pappano, "The Year of the MOOC," *New York Times*, November 2, 2012, http://www.nytimes.com/2012/11/04/education/edlife/massive-open-online-courses-are-multiplying-at-a-rapid-pace.html.

47. A. McAuley, B. Stewart, G. Siemens, and D. Cormier, *The MOOC Model for Digital Practice,* Knowledge Synthesis Grant on the Digital Economy (Ottawa, ON: Social Sciences and Humanities Research Council, 2010), p. 23, http://davecormier.com/edblog/wp-content/uploads/MOOC_Final.pdf.

48. S. Leckart, "The Stanford Education Experiment Could Change Higher Learning Forever," *Wired Magazine*, March 20, 2012, p. 68.

49. C. Rodriguez, "MOOCs and the AI-Stanford Like Courses: Two Successful and Distinct Course Formats for Massive Open Online Courses," *European Journal of Open, Distance and E-Learning* 3 (2012): 7–8, http://www.eurodl.org/materials/contrib/2012/Rodriguez.pdf.

50. MIT, "Harvard and MIT Announce edX," press release, accessed February 14, 2015, http://web.mit.edu/press/2012/mit-harvard-edx-announcement.html.

51. edX Web site, accessed September 11, 2015, https://www.edx.org/schools-partners.

52. S. P. Nanda, "Learn Anytime, Anywhere and Largely for Free," *Live Mint*, August 2, 2013, http://www.livemint.com/Politics/TFqJqM6l4K39Q2gq8JNg5O/Learn-anytime-anywhere-and-largely-for-free-Anant-Agarwal.html.

53. S. Balfour, "Assessing Writing in MOOCs: Automated Essay Scoring and Calibrated Peer Review," *Research and Practice in Assessment* 8, no. 1 (2013): 40–48.

54. M. Freire, A. del Blanco, and B. Fernández-Manjón, "Serious Games as edX MOOC Activities," in *Global Engineering Education Conference (EDUCON)* (New York: Institute of Electrical and Electronics Engineers, 2014), pp. 867–71.

55. A. Kelly, *Disruptor, Distracter, or What? A Policymaker's Guide to Massive Open Online Courses (MOOCs)* (Washington, DC: Bellwether Education Partners, 2014), p. 16, http://www.smarthighered.com/wp-content/uploads/2014/05/MOOC-Final.pdf.

56. TurnItIn Web site, accessed February 14, 2015, http://turnitin.com/en_us/why-turnitin.

57. *A Brief History of the Advanced Placement Program* (New York: College Board, 2003), http://www.collegeboard.com/prod_downloads/about/news_info/ap/ap_history_english.pdf.

58. College-Level Examination Program Web site, accessed August 8, 2015, https://clep.collegeboard.org.

59. R. Rivard, "Udacity Project on 'Pause,'" *Inside Higher Ed*, July 18, 2013, https://www.insidehighered.com/news/2013/07/18/citing-disappointing-student-outcomes-san-jose-state-pauses-work-udacity.

60. C. Straumsheim, "All-MOOC MBA," *Inside Higher Ed*, May, 5, 2015, https://www.insidehighered.com/news/2015/05/05/u-illinois-urbana-champaign-offer-online-mba-through-coursera.

61. "Texas State Promotes Free Frosh (MOOC) Year," *Inside Higher Ed*, September 11, 2015, https://www.insidehighered.com/quicktakes/2015/09/11/texas-state-promotes-free-frosh-mooc-year.

62. S. Brown, "Back to the Future with MOOCs," paper presented at the International Conference on Information Communication Technologies in Education, Crete, Greece, July 4–6, 2013.

63. A. Watters, "The Wrath against Khan: Why Some Educators Are Questioning Khan Academy, *Hacked Education*, July 19, 2011, http://hackeducation.com/2011/07/19/the-wrath-against-khan-why-some-educators-are-questioning-khan-academy.

64. S. Noguchi, "Khan Academy: Amid the Adulation, Some Critical Voices," *San Jose Mercury News*, August 13, 2012, www.mercurynews.com/education/ci_21287389/popular-khan-academy-draws-criticism-first-time.

65. K. Ani, "Khan Academy: The Revolution That Isn't," *Washington Post*, July 23, 2012, http://www.washingtonpost.com/blogs/answer-sheet/post/khan-academy-the-hype-and-the-reality/2012/07/23/gJQAuw4J3W_blog.html.

66. N. DeSantis, "Stanford Professor Gives Up Teaching Position, Hopes to Reach 500,000 Students at Online Start-Up," *Chronicle of Higher Education*, January 23, 2012.

67. R. Rivard, "EdX Rejected," *Inside Higher Ed*, April 19, 2013, www.insidehighered.com/news/2013/04/19/despite-courtship-amherst-decides-shy-away-starmooc-provider.

68. Brown, "Back to the Future with MOOCs."

Chapter 5. Immersive Media

Epigraph: C. Geronimi, W. Jackson, and H. Luske (directors), *Alice in Wonderland*, Walt Disney Company, 1951. Adapted from Lewis Carroll's books *Alice's Adventures in Wonderland* (1865) and *Through the Looking-Glass, and What Alice Found There* (1871).

1. I. Sutherland, "A Head-Mounted Three Dimensional Display," in *Proceedings of the December 9–11, 1968, Fall Joint Computer Conference, Part I* (New York: ACM, 1968), pp. 757–64.

2. D. Staley, *Computers, History and Visualization: How Technology Will Transform Our Understanding of the Past* (Armonk, NY: M. E. Sharpe, 2003), pp. 90–94.

3. C. Dede, "Immersive Interfaces for Engagement and Learning," *Science* 323, no. 66 (2009): 66.

4. Ibid., pp. 66–69.

5. K. Haycock and J. Kemp, "Immersive Learning Environments in Parallel Universes: Learning through Second Life," *School Libraries Worldwide* 14, no. 2 (2008): 89–97.

6. E. Ackermann, "Piaget's Constructivism, Papert's Constructionism: What's the Difference?," in *Constructivism: Uses and Perspectives in Education*, vols. 1 and 2, *Conference Proceedings* (Geneva: Research Center in Education, 2001), pp. 85–94.

7. S. Papert, *Mindstorms: Children, Computers, and Powerful Ideas* (New York: Basic Books, 1993), pp. 7–27.

8. Ackermann, "Piaget's Constructivism," p. 90.

9. Papert, *Mindstorms*, p. 125.

10. K. Pagano, *Immersive Learning: Design for Authentic Practice* (Alexandria, VA: ASTD Press, 2013), p. 16.

11. S. Coleridge, *Biographia Literaria, or Biographical Sketches of My Literary Life and Opinions*, vol. 2 (London: S. Curtis, 1817), p. 2.

12. S. McCloud, *Understanding Comics: The Invisible Art* (New York: HarperCollins, 1993), p. 31.

13. M. Csikszentmihalyi, *Flow: The Psychology of Optimal Experience* (New York: HarperCollins, 1990), p. 3.

14. D. Max, *Every Love Story Is a Ghost Story: A Life of David Foster Wallace* (New York: Penguin, 2012), p. 49.

15. M. Mori, "The Uncanny Valley," *Energy* 7, no. 4 (2012): 34.

16. A. Tinwell, *The Uncanny Valley in Games and Animation* (Boca Raton, FL: Taylor and Francis, 2014), pp. 2–10.

17. J. Clarke and C. Dede, "Design for Scalability: A Case Study of the River City Curriculum," *Journal of Science Education and Technology* 18, no. 4 (2009): 353–65.

18. "Active Worlds," River City Project Web site, accessed February 16, 2015, http://rivercity5.activeworlds.com.

19. W. Au, *The Making of Second Life: Notes from the New World* (New York: HarperCollins, 2008), pp. x–xxi.

20. P. Gray, Linden Labs, personal communication, August 21, 2015.

21. G. McElroy, "The New Doom Game Is Just Titled 'Doom,' Runs on id Tech 6, and More Details," *Polygon*, July 17, 2014, www.polygon.com/2014/7/17/5913883/doom-4-quakecon-reveal.

22. Au, *Making of Second Life*, pp. 16–17.

23. Ibid., pp. 20–37.

24. P. Gray, Linden Labs, personal communication, August 21, 2015.

25. W. Keeney-Kennicutt, personal communication, September 2, 2015.

26. RateMyProfessors Web site, accessed September 23, 2015, www.ratemyprofessors.com/ShowRatings.jsp?tid=234109.

27. W. Keeney-Kennicutt, personal communication, September 1, 2015.

28. "TAMHSC Creates Virtual Campus on Second Life," *vitalRECORD*, January 28, 2011,https://news.tamhsc.edu/?post=tamhsc-creates-virtual-campus-on-second-life.

29. W. Keeney-Kennicutt and Z. Merchant, "Using Virtual Worlds in the General Chemistry Classroom," in *Pedagogic Roles of Animations and Simulations in Chemistry Courses*, vol. 1142, ed. J. Suits and M. Sanger (Washington, DC: American Chemical Society, 2013), pp. 181–204.

30. W. Keeney-Kennicutt and K. Winkelmann, "What Can Students Learn from Virtual Labs?," in *ACS Committee on Computers in Chemical Education Fall Newsletter* (Washington, DC: American Chemical Society, 2013), Paper 9.

31. W. Keeney-Kennicutt, personal communication, September 2, 2015.

32. Ibid.

33. L. Baum, *The Master Key: An Electrical Fairy Tale* (Brooklyn, NY: Bowen-Merrill, 1901), p. 94.

34. D. Pierce, "Google's Cardboard is VR's Gateway Drug," *Wired Magazine*, May 28, 2015, www.wired.com/2015/05/try-google-cardboard.

35. G. Lucas (director), *Star Wars,* Twentieth Century Fox, 1977.

36. B. Barrett, "Microsoft Shows HoloLens' Augmented Reality Is No Gimmick," *Wired Magazine*, May 14, 2015, http://www.wired.com/2015/04/microsoft-build-hololens.

37. A. Walsh, personal communication, October 8, 2015.

38. "Case Western Reserve, Cleveland Clinic Collaborate with Microsoft on 'Earth-Shattering' Mixed-Reality Technology for Education," Case Western Reserve University Web site, accessed April 29, 2015, http://case.edu/hololens.

39. "NASA, Microsoft Collaboration Will Allow Scientists to 'Work on Mars,'" Jet Propulsion Laboratory Web site, accessed January 21, 2015, www.jpl.nasa.gov/news/news.php?feature=4451.

40. D. Minock, personal communication, October 7, 2015.

41. AR Flashcards Web site, accessed October 10, 2015, http://arflashcards.com.

42. "Anatomy 4D," DAQRI Web site, accessed October 11, 2015, http://daqri.com.

43. D. Minock, personal communication, October 7, 2015.

44. Two Guys and Some iPads Web site, accessed October 7, 2015, www.twoguysandsomeipads.com.

45. "Elements 4D," DAQRI Web site, accessed September 25, 2015, http://elements4d.daqri.com.

46. A. Walsh, personal communication, October 8, 2015.

Chapter 6. Making Sense of Media for Learning

1. R. Hackforth, trans., *Plato's Phaedrus* (New York: Cambridge University Press, 1952), p. 157.

2. F. Demana, "Calculators in Mathematics Teaching and Learning: Past, Present, and Future," in *Learning Mathematics for a New Century* (Reston, VA: National Council of Teachers of Mathematics, 2000), http://citeseerx.ist.psu.edu/viewdoc/download?doi=10.1.1.35.7029&rep=rep1&type=pdf.

3. M. McLuhan, *Understanding Media: The Extensions of Man* (New York: McGraw-Hill, 1964).

4. R. Clark, "Reconsidering Research on Learning from Media," *Review of Educational Research* 53, no. 4 (1983): 445.

5. R. Clark, "Media Will Never Influence Learning," *Educational Technology Research and Development* 42, no. 2 (1994): 21.

6. R. Clark, personal communication, October 14, 2015.

7. R. Clark, "Reexamining the Methodology of Research on Media and Technology in Education," *Review of Educational Research* 47, no. 1 (1977): 99–120.

8. Clark, "Media Will Never Influence Learning."

9. R. Kozma, personal communication, October 16, 2015.

10. R. Kozma, "*Will* Media Influence Learning? Reframing the Debate," *Educational Technology Research and Development* 42, no. 2 (1994): 15.

11. R. Kozma, "Learning with Media," *Review* of *Educational Research* 61, no. 2 (1994): 179–212.

12. R. Kozma, personal communication, October 16, 2015.

13. R. Clark, personal communication, October 14, 2015.

14. R. Kozma, personal communication, October 16, 2015.

15. R. Crump, "Correspondence and Class Extension Work in Oklahoma" (doctoral dissertation, Teachers College, Columbia University, 1928).

16. J. Clark, "The Correspondence School—Its Relation to Technical Education and Some of Its Results," *Science* 24, no. 611 (1906): 327–34.

17. Ibid., p. 329.

18. R. Clark, "Bloodletting, Media and Learning," in *The No Significant Difference Phenomenon: A Comparative Research Annotated Bibliography on Technology for Distance Education, As Reported in 355 Research Reports, Summaries and Papers*, ed. T. Russell (Raleigh: North Carolina State University, 1999), pp. 8–11.

19. E. El-Sheikh, "An Architecture for the Generation of Intelligent Tutoring Systems from Reusable Components and Knowledge-Based Systems" (doctoral dissertation, Michigan State University, 2002) and V. Shute, "SMART: Student Modeling Approach for Responsive Tutoring," *User Modeling and User-Adapted Instruction* 5 (1995): 5–10.

20. E. Rushton, *Programmed Learning: The Roanoke Experiment* (Chicago: Encyclopedia Britannica Press, 1965), p. 6.

21. T. Russell, ed., *The No Significant Difference Phenomenon: A Comparative Research Annotated Bibliography on Technology for Distance Education, As Reported in 355 Research Reports, Summaries and Papers* (Raleigh: North Carolina State University, 1999), p. 17.

22. F. Smith, "The Evolution of the Motion Picture: VI. Looking into the Future with Thomas A. Edison," *New York Dramatic Mirror*, July 9, 1913, p. 24.

23. T. Foster, "Instruction by Correspondence," *American Machinist* 29 (1906): 587.

24. "Government Expenditure on Education as % of GDP," World Bank Web site, accessed November 1, 2015, http://data.worldbank.org/indicator/SE.XPD.TOTL.GD.ZS/.

25. E. Brown, "The States That Spend the Most (and the Least) on Education, in One Map," *Washington Post*, June 2, 1015, https://www.washingtonpost.com/news/local/wp/2015/06/02/the-states-that-spend-the-most-and-the-least-on-education-in-one-map.

26. M. Hiatt and C. Stockton, "The Impact of the Flexner Report on the Fate of Medical Schools in North America after 1909," *Journal of the American Medical Association* 291, no. 17 (2003): 2139–140.

27. E. Johanningmeier, "A Nation at Risk and Sputnik," *American Educational History Journal* 37, no. 2 (2010): 347–65.

28. R. Dudney, "When Sputnik Shocked the World," *Air Force Magazine: Journal of the Air Force Association* 90, no. 10 (2007): 42–47.

29. H. Rickover, *Education and Freedom* (New York: Dutton, 1959), p. 6.

30. K. Carey, *The End of College: Creating the Future of Learning and the University of Everywhere* (New York: Riverhead Books, 2015), pp. 130–38.

31. "Results of 4A's 2011 Television Production Cost Survey," American Association of Advertising Agencies Web site, accessed February 17, 2015, www.aaaa.org/news/bulletins/Pages/tvprod_01222013.aspx.

32. N. Minow and N. Minow, "What Are We Learning from Television?," *Change* 8, no. 9 (1976): 49.

33. S. Kanfer, *The Essential Groucho: Writings by, for, and about Groucho Marx* (London: Penguin Books, 2000), p. 207.

34. G. Gordon, "Instructional Television: Yesterday's Magic," in *Instructional Television: Status and Directions*, ed. J. Ackerman and L. Lipsitz (Englewood Cliffs, NJ: Educational Technology, 1977), p. 149.

35. W. Schramm, *Big Media, Little Media: Tools and Technologies for Instruction* (Beverly Hills, CA: Sage, 1977), pp. 26–58.

Index

Note: Page numbers in *italics* indicate figures.